MW00814961

SUPERCONDUCTIVITY, MAGNETISM AND MAGNETS

SUPERCONDUCTIVITY, MAGNETISM AND MAGNETS

LANNIE K. TRAN
EDITOR

Nova Science Publishers, Inc.
New York

For permission to use material from this book please contact us:
Telephone 631-231-7269; Fax 631-231-8175
Web Site: http://www.novapublishers.com

NOTICE TO THE READER

LIBRARY OF CONGRESS CATALOGING-IN-PUBLICATION DATA
Superconductivity, magnetism, and magnets / Lannie K. Tran, editor.
p. cm.
Includes index.
ISBN 1-59454-845-5
. High temperature superconductors. 2. Copper oxide superconductors. 3. High temperature superconductivity. I. Tran, Lannie K.
QC611.98.H54S883 2006
537.6'23--dc22 2005032465

Published by Nova Science Publishers, Inc. ✢ New York

CONTENTS

PREFACE

Superconductivity is the ability of certain materials to conduct electrical current with no resistance and extremely low losses. High temperature superconductors, such as La2-xSrxCuOx (Tc=40K) and YBa2Cu3O7-x (Tc=90K), were discovered in 1987 and have been actively studied since. In spite of an intense, world-wide, research effort during this time, a complete understanding of the copper oxide (cuprate) materials is still lacking. Many fundamental questions are unanswered, particularly the mechanism by which high-Tc superconductivity occurs. More broadly, the cuprates are in a class of solids with strong electron-electron interactions. An understanding of such "strongly correlated" solids is perhaps the major unsolved problem of condensed matter physics with over ten thousand researchers working on this topic. High-Tc superconductors also have significant potential for applications in technologies ranging from electric power generation and transmission to digital electronics. This ability to carry large amounts of current can be applied to electric power devices such as motors and generators, and to electricity transmission in power lines. For example, superconductors can carry as much as 100 times the amount of electricity of ordinary copper or aluminum wires of the same size. Many universities, research institutes and companies are working to develop high-Tc superconductivity applications and considerable progress has been made. This new volume brings together new leading-edge research in the field.

In Chapter 1 the authors develop a simple calculational scheme for thermodynamic properties of superconducting states under magnetic fields. A combination of an approximate analytic solution with a free energy functional in the quasiclassical theory provides a wide use formalism for spatial- averaged thermodynamic properties, and requires a little numerical computation. The theory covers multiband superconductors with various set of singlet and unitary triplet pairings in the presence of an impurity scattering. They demonstrate the application to s-wave, d_{x2-y2-} wave and two-band s-wave pairings, and discuss the validity of the theory comparing with previous numerical studies. Using the formalism, they examine the influence of modulation of gap functions and anisotropy of Fermi velocity to upper critical field and specific heat under the oriented magnetic fields. The validity of the Doppler shift method and its range of applicability in the temperature-field phase diagram is discussed. This result gives a concrete theoretical ground for recent development of the angle-resolved experiments under the oriented magnetic fields.

Chapter 2 is a review of the calculation of surface nucleation of superconductivity in low-temperature type II superconductors with cylindrical symmetry based on the linearized

Ginzburg-Landau (G-L) equation for the order parameter. Two distinct configurations of a spatially inhomogeneous applied magnetic field considered are when the magnetic field is applied either parallel to the axis of the superconducting structure or in the azimuthal direction. In the parallel configuration, the universal temperature-field (ε-f) curves are characterized by flux-entry points at each of which the azimuthal quantum number m decreases by unity. In this case, critical field corresponds strictly to zero axial wave number kz. The quasi-periodicity of the flux-entries increases in f with the increasing degree of the spatial inhomogeneity of the axial applied magnetic field. An increase of the degree of the spatial inhomogeneity of the applied magnetic fiel d is also found to drastically reduce the amplitude of Little-Parks oscillations characteristic of a thin-walled hollow cylinder. The situation is the other way around in the azimuthal configuration, that is $m = 0$ and $kz \neq 0$, although in general kz is a continuous variable. Spatial distributions of the superconducting phases are investigated by studying variations of the order parameter as well as the current density across thick specimen. The magnetic field profile, in particular, in the parallel configuration, can be tailored in such a way as to counteract spatial variations of the superconducting phase particularly across a thick specimen. This has the advantage that in more complex evaluations beyond merely a calculation of the critical field, the specimen may then be described in terms of a *constant* wave function.

In Chapter 3 the authors consider the effects of disorder due to impurities acting as scatterers for the electrons and as pinning centers for vortices, induced by an external magnetic field, in a strongly type-II superconductor. They consider the cases of s-wave and d-wave pairing symmetries. The effect of disorder reflects in the density of states, particularly at low energies. Considering first that no impurities are present but that the vortices are pinned randomly, the density of states as a function of magnetic field increases at low energies, closing the gap in the s-wave case and becoming finite in the d-wave case. Including the effect of the scattering from impurities, placed randomly using the binary alloy model, the density of states and the local density of states (LDOS) have to be solved self-consistently. The results for the LDOS agree qualitatively with experimental results if most vortices are pinned to the impurities. Also, we consider the effect of impurities on the vortex charge.

In recent years hybrid superconducting-ferromagnetic structures have attracted much interest not only due to their importance for basic physics but also for the numerous modern practical applications that could be based on such devices in the near future. In Chapter 4 the authors present results on such hybrid structures (HSs) consisting of CoPt magnetic nanoparticles (MNs) that are randomly embedded at the bottom of high quality Nb superconducting layers. More specifically, in the present work they investigate the influence of the macroscopic magnetic characteristics of the MNs (i.e. saturation and coercive fields, saturation magnetization e.t.c.) on the nucleation of superconductivity in the Nb layer. The effects under discussion are studied by means of magnetization and resistivity measurements in various categories of HSs where MNs that exhibit different magnetic properties are employed. The results demonstrate in a definite way that the MNs can be used at will as an efficient ingredient for the modulation of the superconducting properties of the Nb layer. While in a single superconductor the nucleation of superconductivity depends only on the applied magnetic field and temperature it turns out that in HSs the nucleation of superconductivity can be modulated not only by the parameters mentioned above but also by the magnetic state of the MNs. These facts are clearly revealed by the pronounced magnetic memory effects observed in our data. Furthermore, we find that in our HSs surface superconductivity is strongly enhanced in the high-field regime since on the operational H – T phase diagram the respective boundary line $Hc_3(T)$ exhibits a pronounced upturn when the applied field exceeds the saturation field of MNs. The observed change in slope of $Hc_3(T)$ is determined by the saturation magnetization of MNs. Based on our experimental findings we propose possible practical applications that such HSs could find in the near future. Power applications, where the increasing of current-carrying capability of a superconductor is the main objective, and magnetoresistive memory elements, where the element's resistance should ideally switch between two or three distinct states are discussed.

The mathematical foundation is laid for a relatively new type of magnets for generation of uniform transverse magnetic field in a large volume of space - tilted coil magnets. These consist of concentric nested solenoidal coils with elliptical turns tilted at a certain angle to the central axis and current flowing in opposite directions in the coils tilted at opposite angles, generating a perfectly uniform transverse field. Both superconducting wire-wound and resistive Bitter tilted coils are discussed. An elegant analytical method is used to prove that the wire-wound tilted coils have the ideal distribution of the axial current density, "cosine-theta" plus constant. Magnetic fields are calculated for a tilted Bitter coil magnet using an original exact solution for current density in an elliptical Bitter disk. This solution was obtained for disk shapes bounded by homothetic ellipses. A much simpler analytical solution was found for the case of confocal ellipses. A finite element elastic stress analysis was also conducted and revealed rather moderate, i.e., acceptable, stresses in the coils. Winding technologies for tilted coils are discussed in Chapter 5. Superconducting wire-wound tilted coil magnets may become an alternative to traditional dipole magnets for accelerators, and Bitter tilted coil magnets are attractive for rotation experiments with a large access port perpendicular to the field and for testing of long samples of superconducting wires and cables in high magnetic fields. It is noteworthy that tilted Bitter magnets represent the only option to test long pieces of high field superconductors and other lengthy materials in transverse uniform fields above 15T, which makes such magnets very attractive from a practical standpoint.

AC magnetic susceptibility measurement is perhaps the most useful experimental technique for understanding the magnetic properties of superconductors. This measurement technique includes an in phase (or real) component χ_n' and an imaginary out of phase component χ_n'' (n=1, 2, 3...). The AC magnetic susceptibility considered here is in principle deduced by means of a phase sensitive detector. The phase sensitive detector is operated using a filter to accept the fundamental and higher harmonic susceptibilities and no filtering is used for AC magnetization hysteresis loops. The concept of a critical-state model for superconductor and its use in the interpretation of AC magnetic measurement in terms of critical-current density were introduced by Bean in which critical current density was assumed to be independent of the local magnetic field. Further critical-state models have been proposed, with a different local magnetic field dependence of critical current density. Nevertheless the observed external parameters (AC field amplitude and frequency, dc magnetic field, temperature etc.) of fundamental and higher harmonics can not be described within any critical-state model. In order to explain such dependencies, creep models have been introduced, such as Giant Flux Creep, Vortex Glass and Collective Creep.

In Chapter 6, the AC magnetic susceptibility technique and critical-state models will be briefly reviewed. The experimental AC susceptibility data on some high T_c superconductors will be compared with several theoretical models.

In: Superconductivity, Magnetism and Magnets
Editor: Lannie K. Tran pp. 1-30

ISBN 1-59454-845-5

Chapter 1

THERMODYNAMIC PROPERTIES OF SUPERCONDUCTING STATES UNDER MAGNETIC FIELDS

Hiroaki Kusunose
Department of Physics, Tohoku University, Sendai 980-8578, Japan

Abstract

We develop a simple calculational scheme for thermodynamic properties of superconducting states under magnetic fields. A combination of an approximate analytic solution with a free energy functional in the quasiclassical theory provides a wide use formalism for spatial-averaged thermodynamic properties, and requires a little numerical computation. The theory covers multiband superconductors with various set of singlet and unitary triplet pairings in the presence of an impurity scattering. We demonstrate the application to s-wave, $d_{x^2-y^2}$-wave and two-band s-wave pairings, and discuss the validity of the theory comparing with previous numerical studies. Using the formalism, we examine the influence of modulation of gap functions and anisotropy of Fermi velocity to upper critical field and specific heat under the oriented magnetic fields. The validity of the Doppler shift method and its range of applicability in the temperature-field phase diagram is discussed. This result gives a concrete theoretical ground for recent development of the angle-resolved experiments under the oriented magnetic fields.

1 Introduction

A determination of its gap symmetry has been a long standing issue in a course study of unconventional superconductivity. Because of a modulation of gap amplitude, thermodynamic properties at low temperature follow power-law behaviors, which give a hint of the structure of the gap function[1, 2]. However, this in principle cannot determine the absolute direction of nodes, which has a significant importance in selectively identifying the pairing mechanism.

A magnetic field has been a good probe to investigate a gap symmetry of superconductivity. In particular, the low-energy excitations in the mixed state exhibit unconventional behaviors inherent in different nature of a vortex core, nodal structure and quasiparticle transfer between vortices [3, 4, 5, 6]. In a recent development in multiband superconductors [5, 6, 7, 8, 9, 10, 11], magnetic fields also play an important role. Namely, weak magnetic fields suppress strongly smaller gap of the multiband superconductivity keeping larger gap be unaffected, leading to drastic enhancement in field dependence of the residual density of states [12, 13], the thermal conductivity [14, 15] and so on. Additionally, a powerful method has been developed to measure the modulation of the gap amplitude directly. Namely, field angle dependences in oriented magnetic fields have shown oscillatory behaviors of thermal conductivity[16, 17, 18, 19, 20, 21] and specific heat[22, 23, 24, 25, 26, 27], reflecting low-energy excitations inherent from the gap structure.

To analyze the thermodynamic properties under magnetic fields, various theoretical approaches have been put forward. Needless to say, the Ginzburg-Landau theory is powerful but is restricted to regions near the critical fields[28]. The so-called Doppler shift method, originally proposed by Volovik[29], is used frequently, in which the quasiparticle excitation energy is approximated by taking into account the Doppler shift of the local superconducting current [30, 31, 3, 32]. This method neglects the scattering by vortex cores and the overlap of the core states, which makes the theory to be valid only at low-temperature and low-field regions [33]. Meanwhile, fully numerical calculations for the quasiclassical transport-like equation or the Bogoliubov-de Gennes equation have been performed [34, 35, 36, 37, 38]. Although these approaches are necessary to discuss local structure of the core states, it is not easily applicable for transport properties in the linear-response theory, impurity effect and extension to multiband superconductors with detailed band structure because of intractable aspect of numerical computation.

Experimentally, the field-angle dependences of the thermal conductivity and the specific heat have extensively been studied. An analysis of two experiments, however, is somewhat ambiguous, and results in opposite conclusions at worst[17, 23]. This is because a dominant source of the oscillation differs depending on what fields and temperatures we consider. Apparently, the Doppler shift method can be applied to very low fields and temperatures, and various extrinsic effects come into play in the angle dependences in most of regions in the H-T phase diagram[39, 40, 41, 42, 43]. Therefore, a simple but a reliable theoretical tool to investigate the whole H-T phase diagram is highly desired in order to exclude these extrinsic effects from pure contribution of the gap modulation.

Motivated by these situation, it is useful to arrange a semi-analytic approach to analyze various experimental results and to check their consistency on an equal footing. For weak-coupling superconductors, there exists a reliable approximate analytic solution [44, 45] in the quasiclassical formalism [46, 47]. Although required conditions for the theory seem to be valid near the upper critical field H_{c2}, it is meaningful to extrapolate the theory to lower fields. With understanding of its validity, it becomes very useful to discuss spatial-averaged thermodynamic properties of superconductivity under magnetic fields. The theory can be extended easily to multiband superconductors with various set of singlet and unitary triplet pairings in the presence of an impurity scattering[48]. The realistic band structure can be included in arbitrary direction of magnetic fields. It is also applicable to investigate linear-response coefficients in a similar manner [49, 50, 51, 52, 40], which will be summarized elsewhere.

The paper is organized as follows. In the next section, we introduce the quasiclassical equations in a general form, then we give explicit expressions of the transport-like equation and the free energy functional for singlet and unitary triplet pairings. In §3 we explain the approximate analytic solution for quasiclassical equations in the presence of the impurity scattering. We express useful formulas for basic thermodynamic quantities. Then, we extend the theory to multiband superconductors. The application of the method to s-wave, $d_{x^2-y^2}$ and two-band s-wave pairings and the comparison with previous numerical solutions are given in §4. The magnetization curve is also discussed with a comparison with the appropriate Ginzburg-Landau result. In §5, we discuss the influence of the gap modulation and the Fermi velocity anisotropy in H_{c2} and C taking Sr_2RuO_4 as an example. The last section summarizes the paper.

2 Quasiclassical Equations

The microscopic Green's function contains all the information about the single-particle properties. In most cases, however, states far from the Fermi energy or rapid oscillations inherent in the Fermi surface play little role to superconducting properties. The key idea of the quasiclassical theory is that those irrelevant states are integrated out from the beginning in the Gor'kov formalism [46, 47, 50, 51]. Then, the basic equations of the quasiclassical theory give a complete description of the superconducting properties on a coarse-grained scale. In this section, we first review the derivation of the standard quasiclassical equations under magnetic fields, and we introduce necessary notations through this step. Next, we restrict ourselves to the cases of the singlet and the unitary triplet pairings, and then we construct a free energy functional as a generating functional for the quasiclassical formalism.

2.1 Derivation of Quasiclassical Equations

Let us start with the mean-field (MF) Hamiltonian in the 4×4 Nambu-Gor'kov space with a spatial inhomogeneity ($c = \hbar = k_{\rm B} = 1$ hereafter),

$$H_{\rm MF} = \frac{1}{2}\int d\boldsymbol{x}_1 d\boldsymbol{x}_2 \boldsymbol{\Psi}^\dagger(\boldsymbol{x}_1)\left(\hat{K}^{12} + \hat{\Delta}^{12}\right)\hat{\rho}^z \boldsymbol{\Psi}(\boldsymbol{x}_2), \tag{1}$$

where the creation-annihilation field operators at the position $\boldsymbol{x}_s$ are composed as $\boldsymbol{\Psi}^\dagger(\boldsymbol{x}_s) = [\psi_\uparrow^\dagger(\boldsymbol{x}_s), \psi_\downarrow^\dagger(\boldsymbol{x}_s), \psi_\uparrow(\boldsymbol{x}_s), \psi_\downarrow(\boldsymbol{x}_s)]$. An operator with a hat acts on the Nambu-Gor'kov space, and $\hat{\boldsymbol{\rho}}$ ($\hat{\boldsymbol{\sigma}}$) is the vector composed of Pauli matrices on the 2×2 particle-hole (spin) space remaining the other degrees of freedom unchanged. The unit matrix is denoted by $\hat{1}$. The kinetic-energy matrix is then given by

$$\hat{K}^{12} = \delta(\boldsymbol{x}_1 - \boldsymbol{x}_2)\xi\left(-i\boldsymbol{\nabla}_2\hat{\rho}^z - e\boldsymbol{A}_2\hat{1}\right), \tag{2}$$

with a single-particle energy operator $\xi(\boldsymbol{p})$, where $\boldsymbol{A}_s$ is the vector potential at $\boldsymbol{x}_s$ and $e < 0$ is the electron charge. The superconducting gap matrix is given by

$$\hat{\Delta}^{12} = \begin{bmatrix} 0 & \Delta(\boldsymbol{x}_1, \boldsymbol{x}_2) \\ -\Delta^\dagger(\boldsymbol{x}_2, \boldsymbol{x}_1) & 0 \end{bmatrix}, \tag{3}$$

and its component is defined in terms of the two-body interaction $V_{\alpha\beta\alpha'\beta'}(\boldsymbol{x}_1, \boldsymbol{x}_2)$ as

$$\Delta_{\alpha\beta}(\boldsymbol{x}_1, \boldsymbol{x}_2) = -\sum_{\alpha'\beta'} V_{\alpha\beta\alpha'\beta'}(\boldsymbol{x}_1, \boldsymbol{x}_2)\langle\psi_{\beta'}(\boldsymbol{x}_2)\psi_{\alpha'}(\boldsymbol{x}_1)\rangle. \tag{4}$$

Corresponding to the MF Hamiltonian, the Nambu-Gor'kov equation reads

$$\int dy \left[-\delta(x_1 - y)\hat{\rho}^z \frac{\partial}{\partial\tau_1} - \delta(\tau_1 - \tau)\left(\hat{K}^{1y} + \hat{\Delta}^{1y}\right) - \hat{\Sigma}^{1y}(\tau_1 - \tau)\hat{\rho}^z\right] \hat{\rho}^z \hat{G}(\boldsymbol{y}, \boldsymbol{x}_2; \tau_1 - \tau_2) = \tag{5}$$

for the imaginary-time Green's function

$$\hat{G}(\boldsymbol{x}_1, \boldsymbol{x}_2; \tau_1 - \tau_2) = -\langle T_\tau \Psi(\boldsymbol{x}_1, \tau_1)\Psi^\dagger(\boldsymbol{x}_2, \tau_2)\rangle, \tag{6}$$

where we have used $x_s = (\boldsymbol{x}_s, \tau_s)$ and $y = (\boldsymbol{y}, \tau)$. Here the operator T_τ arranges the field operators in ascending order of the imaginary time $0 < \tau < 1/T$. The self-energy $\hat{\Sigma}$ has been taken into account to treat the impurity scattering in subsequent discussion.

In order to integrate out the rapid oscillations, we introduce the center-of-mass coordinate, $\boldsymbol{R} = (\boldsymbol{x}_1 + \boldsymbol{x}_2)/2$ and the relative coordinate, $\boldsymbol{r} = \boldsymbol{x}_1 - \boldsymbol{x}_2$, and perform the Fourier transformation in the latter according to

$$f_{\boldsymbol{R}}(\boldsymbol{k}; i\omega_n) = \int d\boldsymbol{r} \int_0^{1/T} d\tau f(\boldsymbol{x}_1, \boldsymbol{x}_2; \tau)\, e^{-i(\boldsymbol{k}\cdot\boldsymbol{r} - \omega_n\tau)}, \tag{7}$$

where $\omega_n = (2n+1)\pi T$ is the fermionic Matsubara frequency. A product of functions, $h^{12}(\tau_1 - \tau_2) = \int d\boldsymbol{y} \int d\tau f^{1y}(\tau_1 - \tau) g^{y2}(\tau - \tau_2)$, is then transformed into so-called "circle-product" (see Appendix A) [53],

$$h_{\boldsymbol{R}}(\boldsymbol{k}; i\omega_n) = \exp\left[\frac{1}{2i}(\boldsymbol{\nabla}_{\boldsymbol{k}'} \cdot \boldsymbol{\nabla}_{\boldsymbol{R}} - \boldsymbol{\nabla}_{\boldsymbol{k}} \cdot \boldsymbol{\nabla}_{\boldsymbol{R}'})\right] \times f_{\boldsymbol{R}}(\boldsymbol{k}; i\omega_n) g_{\boldsymbol{R}'}(\boldsymbol{k}'; i\omega_n)\Big|_{\boldsymbol{k}'=\boldsymbol{k}, \boldsymbol{R}'=\boldsymbol{R}}. \tag{8}$$

In translationally invariant systems, the internal magnetic field, $\boldsymbol{B} = \boldsymbol{\nabla} \times \boldsymbol{A}$ is only the source of slowly varying field dependence. In the lowest order in $\boldsymbol{\nabla}_{\boldsymbol{R}}$, which is necessary in this paper, the circle-product is given simply by the product of each Fourier transformations, $h_{\boldsymbol{R}}(\boldsymbol{k}) \sim f_{\boldsymbol{R}}(\boldsymbol{k}) g_{\boldsymbol{R}}(\boldsymbol{k})$. Noting the transformation,

$$\xi(-i\boldsymbol{\nabla}_s\hat{\rho}^z - e\boldsymbol{A}_s) \to \xi_{\boldsymbol{k}} + \frac{i}{2}\boldsymbol{v}(\hat{\boldsymbol{k}}) \cdot \left[(-1)^s \boldsymbol{\nabla}_{\boldsymbol{R}} + 2ie\boldsymbol{A}_{\boldsymbol{R}}\hat{\rho}^z\right], \tag{9}$$

in the leading order, we obtain the Nambu-Gor'kov equation as

$$\left[\left(i\omega_n + e\boldsymbol{v}(\hat{\boldsymbol{k}}) \cdot \boldsymbol{A}_{\boldsymbol{R}}\right)\hat{\rho}^z - \left(\xi_{\boldsymbol{k}} - \frac{i}{2}\boldsymbol{v}(\hat{\boldsymbol{k}}) \cdot \boldsymbol{\nabla}_{\boldsymbol{R}}\right)\hat{1} - \hat{\Delta}_{\boldsymbol{R}}(\hat{\boldsymbol{k}}) - \hat{\sigma}_{\boldsymbol{R}}(\hat{\boldsymbol{k}}; i\omega_n)\right] \hat{\rho}^z \hat{G}_{\boldsymbol{R}}(\boldsymbol{k}; i\omega_n) \tag{10}$$

where $\hat{\boldsymbol{k}} = \boldsymbol{k}_{\rm F}/|\boldsymbol{k}_{\rm F}|$ is the unit wave vector at the Fermi surface and $\boldsymbol{v}(\hat{\boldsymbol{k}}) = \boldsymbol{\nabla}_{\boldsymbol{k}}\xi(\boldsymbol{k}_{\rm F})$ is the Fermi velocity. Here we have considered that the self-energy in question has weak momentum dependence with respect to $k_{\rm F}$, hence we have denoted as $\hat{\sigma}_{\boldsymbol{R}}(\hat{\boldsymbol{k}}; i\omega_n) = \hat{\Sigma}(\boldsymbol{k}; i\omega_n)|_{\boldsymbol{k}=\boldsymbol{k}_{\rm F}}\hat{\rho}^z$.

To obtain a closed quasiclassical theory, we must construct equations for a quasiclassical propagator, $\hat{g} \sim \int d\xi \hat{G}$. However, a simple integration of eq. (10) results in unphysical divergences due to $\xi_{\hat{\boldsymbol{k}}}\hat{G}$ term in the left-hand side and $\hat{1}$ term in the right-hand side. We get rid of this inconvenience with the help of the "left-right subtraction trick", which transforms the Dyson's equation into transport-like equations for $\hat{g}$. The trick is to eliminate those divergent terms by subtracting the right-hand Dyson equation, in which operators in the brace of eq. (5) act on the second argument of $\hat{G}$ from the right. A similar argument leading to eq. (10) gives the right-hand equation,

$$\hat{\rho}^z \hat{G}_{\boldsymbol{R}}(\boldsymbol{k}; i\omega_n) \left[\left(i\omega_n + e\boldsymbol{v}(\hat{\boldsymbol{k}}) \cdot \boldsymbol{A}_{\boldsymbol{R}} \right) \hat{\rho}^z - \left(\xi_{\boldsymbol{k}} + \frac{i}{2} \boldsymbol{v}(\hat{\boldsymbol{k}}) \cdot \overleftarrow{\boldsymbol{\nabla}}_{\boldsymbol{R}} \right) \hat{1} - \hat{\Delta}_{\boldsymbol{R}}(\hat{\boldsymbol{k}}) - \hat{\sigma}_{\boldsymbol{R}}(\hat{\boldsymbol{k}}; i\omega_n) \right] = \hat{1}. \tag{11}$$

The unphysical terms then cancel, and the terms left can be ξ-integrated to yield the following transport-like equations,

$$\left[\left(i\omega_n + e\boldsymbol{v}(\hat{\boldsymbol{k}}) \cdot \boldsymbol{A}_{\boldsymbol{R}} \right) \hat{\rho}^z - \hat{\Delta}_{\boldsymbol{R}}(\hat{\boldsymbol{k}}) - \hat{\sigma}_{\boldsymbol{R}}(\hat{\boldsymbol{k}}; i\omega_n) \;,\; \hat{g}_{\boldsymbol{R}}(\hat{\boldsymbol{k}}; i\omega_n) \right] + i\boldsymbol{v}(\hat{\boldsymbol{k}}) \cdot \boldsymbol{\nabla}_{\boldsymbol{R}} \hat{g}_{\boldsymbol{R}}(\hat{\boldsymbol{k}}; i\omega_n) = 0, \tag{12}$$

where we have defined the quasiclassical propagators as

$$\hat{g}_{\boldsymbol{R}}(\hat{\boldsymbol{k}}; i\omega_n) = \int \frac{d\xi}{\pi} \hat{\rho}^z \hat{G}_{\boldsymbol{R}}(\boldsymbol{k}; i\omega_n). \tag{13}$$

At the subtraction step, a normalization of $\hat{g}$ is lost. The appropriate normalization is known as $\hat{g}^2_{\boldsymbol{R}}(\hat{\boldsymbol{k}}; i\omega_n) = -\hat{1}$, which can be confirmed explicitly in a homogeneous case without the internal field $\boldsymbol{B}_{\boldsymbol{R}}$ [53].

To complete the quasiclassical formalism, we give self-consistent equations for the gap and the superconducting current as,

$$\Delta_{\boldsymbol{R}\alpha\beta}(\hat{\boldsymbol{k}}) = -\pi N_0 T \sum_n \sum_{\alpha'\beta'} \times \left\langle V_{\alpha\beta\alpha'\beta'}(\hat{\boldsymbol{k}}, \hat{\boldsymbol{k}}') \hat{g}^{12}_{\boldsymbol{R}\alpha'\beta'}(\hat{\boldsymbol{k}}'; i\omega_n) \right\rangle_{\hat{\boldsymbol{k}}'}, \tag{14}$$

$$\boldsymbol{\nabla}_{\boldsymbol{R}} \times (\boldsymbol{B}_{\boldsymbol{R}} - \boldsymbol{H}_{\boldsymbol{R}}) = 4e\pi^2 N_0 T \sum_n \sum_\alpha \left\langle \boldsymbol{v}(\hat{\boldsymbol{k}}) \hat{g}^{11}_{\boldsymbol{R}\alpha\alpha}(\hat{\boldsymbol{k}}; i\omega_n) \right\rangle, \tag{15}$$

where $\boldsymbol{H}_{\boldsymbol{R}}$ is the external magnetic field and $\hat{g}^{ij}$ is understood as the (i,j)-element in the particle-hole space. The bracket $\langle \cdots \rangle = \int d\hat{\boldsymbol{k}} v^{-1}(\hat{\boldsymbol{k}}) \cdots / \int d\hat{\boldsymbol{k}} v^{-1}(\hat{\boldsymbol{k}})$ represents the angular average over the Fermi surface and N_0 is the density of states (DOS) per spin at the Fermi energy.

2.2 Cases of Singlet and Unitary Triplet Pairings and Generating Functional of Quasiclassical Theory

In this subsection, we give only expressions for a unitary triplet pairing. For a singlet pairing, vector quantities should be replaced by scalar ones and a unit matrix is used instead of $\boldsymbol{\sigma}$.

In the case of a unitary triplet pairing, the spin structure of the gap is decomposed as

$$\Delta_{\boldsymbol{R}\alpha\beta}(\hat{\boldsymbol{k}}) = \boldsymbol{\Delta}_{\boldsymbol{R}}(\hat{\boldsymbol{k}}) \cdot (\boldsymbol{\sigma} i\sigma^y)_{\alpha\beta}, \tag{16}$$

with $i(\mathbf{\Delta} \times \mathbf{\Delta}^*) = 0$ [2]. The quasiclassical propagator also has the form,

$$\hat{g}_{\boldsymbol{R}} = \begin{bmatrix} -ig_{\boldsymbol{R}}\delta_{\alpha\beta} & \boldsymbol{f}_{\boldsymbol{R}} \cdot (\boldsymbol{\sigma} i\sigma^y)_{\alpha\beta} \\ -\boldsymbol{f}^\dagger_{\boldsymbol{R}} \cdot (\boldsymbol{\sigma} i\sigma^y)^\dagger_{\alpha\beta} & ig_{\boldsymbol{R}}\delta_{\alpha\beta} \end{bmatrix}. \tag{17}$$

By definitions (6) and (13), the normal and the anomalous propagators have the symmetry [50],

$$g_{\boldsymbol{R}}(\hat{\boldsymbol{k}}; i\omega_n) = -g_{\boldsymbol{R}}(-\hat{\boldsymbol{k}}; -i\omega_n) = g^*_{\boldsymbol{R}}(-\hat{\boldsymbol{k}}; i\omega_n), \tag{18}$$

and

$$\boldsymbol{f}_{\boldsymbol{R}}(\hat{\boldsymbol{k}}; i\omega_n) = \mp \boldsymbol{f}_{\boldsymbol{R}}(-\hat{\boldsymbol{k}}; -i\omega_n) = \mp \boldsymbol{f}^{\dagger *}_{\boldsymbol{R}}(-\hat{\boldsymbol{k}}; i\omega_n), \tag{19}$$

where the upper (lower) sign is applied for the triplet (singlet) state. Note that the gap function satisfies $\mathbf{\Delta}_{\boldsymbol{R}}(\hat{\boldsymbol{k}}) = \mp \mathbf{\Delta}_{\boldsymbol{R}}(-\hat{\boldsymbol{k}})$ similar to (19). According to these relations, we consider only $\omega_n > 0$ hereafter without loss of generality.

At this point we explicitly take into account an effect of impurity scattering. Since an impurity potential is short range as much shorter than length scale of $\boldsymbol{R}$ dependence, we merely follow the standard t-matrix theory to treat non-magnetic impurities [54, 55, 56]. Decomposing the impurity self-energy similar to eq. (17),

$$\hat{\sigma}_{\boldsymbol{R}} = \begin{bmatrix} -i\sigma^{(n)}_{\boldsymbol{R}}\delta_{\alpha\beta} & \boldsymbol{\sigma}^{(a)}_{\boldsymbol{R}} \cdot (\boldsymbol{\sigma} i\sigma^y)_{\alpha\beta} \\ -\boldsymbol{\sigma}^{(a)\dagger}_{\boldsymbol{R}} \cdot (\boldsymbol{\sigma} i\sigma^y)^\dagger_{\alpha\beta} & i\sigma^{(n)}_{\boldsymbol{R}}\delta_{\alpha\beta} \end{bmatrix}, \tag{20}$$

we obtain

$$\sigma^{(n)}_{\boldsymbol{R}}(i\omega_n) = \frac{1}{2\tau_0} \frac{\langle g_{\boldsymbol{R}} \rangle}{\cos^2\delta + D_{\boldsymbol{R}}\sin^2\delta}, \tag{21}$$

$$\boldsymbol{\sigma}^{(a)}_{\boldsymbol{R}}(i\omega_n) = \frac{1}{2\tau_0} \frac{\langle \boldsymbol{f}_{\boldsymbol{R}} \rangle}{\cos^2\delta + D_{\boldsymbol{R}}\sin^2\delta}, \tag{22}$$

with $D_{\boldsymbol{R}} = \langle g_{\boldsymbol{R}} \rangle^2 + \langle \boldsymbol{f}_{\boldsymbol{R}} \rangle \cdot \langle \boldsymbol{f}^\dagger_{\boldsymbol{R}} \rangle$, where τ_0 is the quasiparticle lifetime in the normal state and δ is the impurity phase shift (e.g. $\delta = 0$ corresponds to the Born approximation and $\delta = \pi/2$ is the unitarity limit). Here $\sigma^{(n)}_{\boldsymbol{R}}$ and $\boldsymbol{\sigma}^{(a)}_{\boldsymbol{R}}$ satisfy the symmetry relation similar to (18) and (19), respectively. For convenience, we introduce the renormalized frequency and gap function as

$$\tilde{\omega}_{n\boldsymbol{R}}(i\omega_n) = \omega_n + \sigma^{(n)}_{\boldsymbol{R}}, \tag{23}$$

$$\tilde{\mathbf{\Delta}}_{\boldsymbol{R}}(\hat{\boldsymbol{k}}; i\omega_n) = \mathbf{\Delta}_{\boldsymbol{R}}(\hat{\boldsymbol{k}}) + \boldsymbol{\sigma}^{(a)}_{\boldsymbol{R}}, \tag{24}$$

then the self-energies can be absorbed formally into the "tilde" quantities in quasiclassical equations.

The explicit expression of eq. (12) is then given by

$$\begin{aligned} &\tilde{\mathcal{L}}_+ \boldsymbol{f}_{\boldsymbol{R}} = g_{\boldsymbol{R}} \tilde{\mathbf{\Delta}}_{\boldsymbol{R}}(\hat{\boldsymbol{k}}; i\omega_n), \qquad \tilde{\mathcal{L}}_- \boldsymbol{f}^\dagger_{\boldsymbol{R}} = g_{\boldsymbol{R}} \tilde{\mathbf{\Delta}}^\dagger_{\boldsymbol{R}}(\hat{\boldsymbol{k}}; i\omega_n), \\ &\boldsymbol{v}(\hat{\boldsymbol{k}}) \cdot \boldsymbol{\nabla}_{\boldsymbol{R}} g_{\boldsymbol{R}} = \tilde{\mathbf{\Delta}}^\dagger_{\boldsymbol{R}}(\hat{\boldsymbol{k}}; i\omega_n) \cdot \boldsymbol{f}_{\boldsymbol{R}} - \boldsymbol{f}^\dagger_{\boldsymbol{R}} \cdot \tilde{\mathbf{\Delta}}_{\boldsymbol{R}}(\hat{\boldsymbol{k}}; i\omega_n), \end{aligned} \tag{25}$$

with the normalization, $g_{\boldsymbol{R}}^2 + \boldsymbol{f}_{\boldsymbol{R}} \cdot \boldsymbol{f}_{\boldsymbol{R}}^\dagger = 1$, where we have introduced

$$\mathcal{L}_\pm(\hat{\boldsymbol{k}}, \boldsymbol{R}; i\omega_n) = \omega_n \pm \frac{1}{2}\boldsymbol{v}(\hat{\boldsymbol{k}}) \cdot [\boldsymbol{\nabla}_{\boldsymbol{R}} \mp 2ie\boldsymbol{A}_{\boldsymbol{R}}], \tag{26}$$

and $\tilde{\mathcal{L}}_\pm = \mathcal{L}_\pm(\omega_n \to \tilde{\omega}_{n\boldsymbol{R}})$. Note that one of three equations in (25) is independent due to the symmetry relations (18) and (19) and the normalization condition. In the case of $\boldsymbol{B}_{\boldsymbol{R}} = 0$, the formal solutions of eq. (25) are given by

$$g = \left[1 + \frac{\tilde{\boldsymbol{\Delta}}(\hat{\boldsymbol{k}}; i\omega_n) \cdot \tilde{\boldsymbol{\Delta}}^\dagger(\hat{\boldsymbol{k}}; i\omega_n)}{\tilde{\omega}_n^2}\right]^{-1/2}, \tag{27}$$

$$\boldsymbol{f} = \frac{\tilde{\boldsymbol{\Delta}}(\hat{\boldsymbol{k}}; i\omega_n)}{\tilde{\omega}_n} g, \qquad \boldsymbol{f}^\dagger = \frac{\tilde{\boldsymbol{\Delta}}^\dagger(\hat{\boldsymbol{k}}; i\omega_n)}{\tilde{\omega}_n} g. \tag{28}$$

The explicit solutions can be obtained by solving the self-consistent equations with eqs. (21)–(24).

The interaction leading to the triplet (singlet) pairing has the form

$$V_{\alpha\beta\alpha'\beta'}(\hat{\boldsymbol{k}}, \hat{\boldsymbol{k}}') = P^\pm_{\alpha\beta\alpha'\beta'} V(\hat{\boldsymbol{k}}, \hat{\boldsymbol{k}}'), \tag{29}$$

where $P^\pm_{\alpha\beta\alpha'\beta'} = (\delta_{\alpha\alpha'}\delta_{\beta\beta'} \pm \delta_{\alpha\beta'}\delta_{\beta\alpha'})/2$ is the projection operator to the triplet (singlet) state. The interaction for the triplet (singlet) pairing is odd (even) against $\hat{\boldsymbol{k}} \to -\hat{\boldsymbol{k}}$ or $\hat{\boldsymbol{k}}' \to -\hat{\boldsymbol{k}}'$. Then we classify the interaction and the gap function by irreducible representations of the point-group [2]. Considering a certain irreducible representation Γ with the highest transition temperature, T_c, we can write $V(\hat{\boldsymbol{k}}, \hat{\boldsymbol{k}}')$ in a separable form,

$$V(\hat{\boldsymbol{k}}, \hat{\boldsymbol{k}}') = -V \sum_\gamma \varphi_\gamma(\hat{\boldsymbol{k}}) \varphi_\gamma^*(\hat{\boldsymbol{k}}'), \tag{30}$$

with $V > 0$, where the basis functions satisfy the orthonormal relation, $\left\langle \varphi_\gamma^*(\hat{\boldsymbol{k}}) \varphi_{\gamma'}(\hat{\boldsymbol{k}}) \right\rangle = \delta_{\gamma\gamma'}$. Suppose $\boldsymbol{\Delta}_{\boldsymbol{R}}(\hat{\boldsymbol{k}}) = \boldsymbol{\Delta}_{\boldsymbol{R}} \varphi_\gamma(\hat{\boldsymbol{k}})$, the gap equation is rewritten as

$$V^{-1}\boldsymbol{\Delta}_{\boldsymbol{R}} = \pi T N_0 \sum_n \left\langle \varphi_\gamma^*(\hat{\boldsymbol{k}}) \boldsymbol{f}_{\boldsymbol{R}}(\hat{\boldsymbol{k}}; i\omega_n) \right\rangle. \tag{31}$$

Equations (15), (25) and (31) constitute the quasiclassical formalism to describe superconducting states under magnetic fields. Alternatively the system of equations can be derived as the saddle point of the following generating functional per unit volume (measured from the energy in the normal state) [46],

$$\Omega_{\rm SN}\left[\boldsymbol{\Delta}_{\boldsymbol{R}}, \boldsymbol{\Delta}^*_{\boldsymbol{R}}, \boldsymbol{f}_{\boldsymbol{R}}, \boldsymbol{f}^\dagger_{\boldsymbol{R}}, \boldsymbol{A}_{\boldsymbol{R}}\right] = \int d\boldsymbol{R} \left[\frac{(\boldsymbol{B}_{\boldsymbol{R}} - \boldsymbol{H}_{\boldsymbol{R}})^2}{8\pi} + V^{-1}|\boldsymbol{\Delta}_{\boldsymbol{R}}|^2 - 2\pi T N_0 \sum_{n \geq 0} \left(\left\langle I_{\boldsymbol{R}}(\hat{\boldsymbol{k}}; i\omega_n) \right\rangle + I'_{\boldsymbol{R}}(i\omega_n)\right)\right], \tag{32}$$

where

$$I_{\boldsymbol{R}}(\hat{\boldsymbol{k}}; i\omega_n) = \varphi_\gamma^*(\hat{\boldsymbol{k}}) \boldsymbol{\Delta}^*_{\boldsymbol{R}} \cdot \boldsymbol{f}_{\boldsymbol{R}} + \boldsymbol{f}^\dagger_{\boldsymbol{R}} \cdot \boldsymbol{\Delta}_{\boldsymbol{R}} \varphi_\gamma(\hat{\boldsymbol{k}}) - \frac{1}{1 + g_{\boldsymbol{R}}} \left[\boldsymbol{f}^\dagger_{\boldsymbol{R}} \cdot \mathcal{L}_+ \boldsymbol{f}_{\boldsymbol{R}} + \boldsymbol{f}_{\boldsymbol{R}} \cdot \mathcal{L}_- \boldsymbol{f}^\dagger_{\boldsymbol{R}}\right] \tag{33}$$

and

$$I'_{\boldsymbol{R}}(i\omega_n) = \frac{1}{2\tau_0 \sin^2\delta} \ln[\cos^2\delta + D_{\boldsymbol{R}} \sin^2\delta]. \tag{34}$$

In eq. (32) the summation over the Matsubara frequencies must be cut off at very large but a finite frequency. This cut-off procedure can be avoided by a prescription,

$$V^{-1} \to N_0 \left[\ln\left(\frac{T}{T_c}\right) + 2\pi T \sum_{n=0}^{\infty} \frac{1}{\omega_n} \right], \tag{35}$$

namely, the cut-off, ω_c, is absorbed in T_c and the second term cancels the contribution from the third term in eq. (32) in the limit of $n \to \infty$. Note that $T_c = (2\omega_c e^{\gamma}/\pi) e^{-1/N_0 V}$, γ being Euler's constant, is the transition temperature at $\boldsymbol{H} = 0$ without the impurity scattering. In practical numerical computations, the summation is simply cut off at $n_c = \omega_c / 2\pi Tc$, where $n_c \sim 50$–100 is appropriate depending on the lowest temperature in problem.

If the quasiclassical equations, eq. (25) are solved separately for given $\boldsymbol{\Delta}_{\boldsymbol{R}}$, $\boldsymbol{\Delta}^*_{\boldsymbol{R}}$ and $\boldsymbol{A}_{\boldsymbol{R}}$, eq. (33) becomes

$$I_R = \boldsymbol{\Delta}^*_R(\hat{\boldsymbol{k}}) \cdot \boldsymbol{f}_R + \boldsymbol{f}^\dagger_R \cdot \boldsymbol{\Delta}_R(\hat{\boldsymbol{k}}) + 2\sigma^{(n)}_R (1 - g_R) - \frac{g_R}{1 + g_R} \left[\tilde{\boldsymbol{\Delta}}^\dagger_R(\hat{\boldsymbol{k}}; i\omega_n) \cdot \boldsymbol{f}_R + \boldsymbol{f}^\dagger_R \cdot \tilde{\boldsymbol{\Delta}}_R(\hat{\boldsymbol{k}}; i\omega_n) \right], \tag{36}$$

and $\Omega_{\rm SN}$ coincides with the physical free-energy functional of $\boldsymbol{\Delta}_{\boldsymbol{R}}$, $\boldsymbol{\Delta}^*_{\boldsymbol{R}}$ and $\boldsymbol{A}_{\boldsymbol{R}}$. In the next section, we show that a simple approximation provides an approximate analytic solution of eq. (25).

3 Approximate Analytic Solution

In the previous section, we have derived the set of the quasiclassical equations as the saddle point of the free energy functional. Basically, a minimum of the free energy functional provides a complete description for thermodynamic properties of superconducting states under magnetic fields. In this section, however, we show that a simple but a reliable approximation drastically simplifies the problem. In the following subsection, we show that a kind of "mean-field" approach leads to an approximate analytic solution for quasiclassical equations (25). Then we discuss self-consistent equations for the impurity scattering self-energy. Lastly, expressions of various thermodynamic quantities are given.

3.1 BPT Approximation

The approximation used here was first proposed in the Gor'kov formalism by Brandt, Pesch and Tewordt (BPT) [44], and later Pesch obtained analytic solutions in the quasiclassical formalism [45]. Although BPT initially aimed to describe superconducting states near the upper critical field H_{c2}, it is meaningful to extrapolate the method to lower fields. In this direction, Dahm and co-workers have made a comparison with the numerical solution of quasiclassical equations [33]. However, in numerical calculation they assumed the structure of the vortex cores and did not solve whole equations self-consistently. A reasonable comparison of the BPT approximation with full numerical solutions will be discussed after the approximation and the formulation are introduced.

In the BPT theory the internal field $\boldsymbol{B_R}$ is approximated by its spatial average $\boldsymbol{B}$. The average will be determined variationally in the present formalism. Hereafter a spatial average of a quantity $X_{\boldsymbol{R}}$ is denoted by X or $\overline{X_{\boldsymbol{R}}}$. An Abrikosov solution[28] is then used for vortex lattice structures for all field range above the lower critical field. When the internal magnetic field $\boldsymbol{B}$ sets parallel to z axis, we write the Abrikosov solution with the Landau gauge $\boldsymbol{A_R} = (0, Bx, 0)$ as

$$\Delta_{\boldsymbol{R}} = \sum_n C_n e^{-ip_n y'} \Phi_0 \left[(x' - \Lambda^2 p_n)/\Lambda\right], \tag{37}$$

where $p_n = 2\pi n/\beta$, β being the lattice constant in the y direction and $\Lambda = (2|e|B)^{-1/2}$ is of the order of the lattice spacing of the vortex lattice. We have taken into account anisotropy of the Fermi velocity in the x-y plane by transferring the coordinates as

$$x = \frac{\xi_x}{\xi_{\text{eff}}} x', \qquad y = \frac{\xi_y}{\xi_{\text{eff}}} y', \tag{38}$$

where ξ_i $(i = x, y)$ is the i-component of the coherence length proportional to the square-root of $\langle v_i^2(\hat{\boldsymbol{k}})\rangle$ and $\xi_{\text{eff}} = \sqrt{\xi_x \xi_y}$ sets equal to Λ. Here $\Phi_0[x] = \exp(-x^2/2)$ is the lowest eigenfunction of a harmonic oscillator and the periodicity of the coefficients C_n specifies the type of vortex lattice. We consider here a scalar gap function for notational simplicity. For the vector gap in the case of the unitary triplet state, the argument can be applied separately to each component.

In addition to the above approximations, we expect that the uniform component $g = \overline{g_{\boldsymbol{R}}}$ could describe a global suppression of the gap amplitude as an average. This is because the higher Fourier $\boldsymbol{K}$ components of $g_{\boldsymbol{R}}$ decrease rapidly as $\exp(-\Lambda^2 K^2)$ and the normal propagator $g_{\boldsymbol{R}}$ does not contain an important phase variation due to vortices [44]. On the other hand, the exact spatial form of $\boldsymbol{\Delta_R}$ due to its phase winding around vortices will be taken into account in determining the anomalous propagator $\boldsymbol{f_R}$ [45].

Within the above approximation, we can solve analytically the quasiclassical equations (25). It is sufficient to solve the first two equations in (25). The argument here essentially follows the discussion by Houghton and Vekhter for s-wave superconductors with the operator formalism [52]. Within the above approximations the anomalous propagator is given by

$$\boldsymbol{f_R} = g\varphi_\gamma(\hat{\boldsymbol{k}})\eta^{(a)}(i\omega_n)\tilde{\mathcal{L}}_+^{-1}\boldsymbol{\Delta_R}. \tag{39}$$

Here we have assumed that the renormalized gap function has the same $\boldsymbol{R}$ dependence as the bare gap $\boldsymbol{\Delta_R}$ and the renormalization factors of the frequency and the gap are independent of $\boldsymbol{R}$, i.e.,

$$\tilde{\omega}_n = \eta^{(n)}(i\omega_n)\omega_n, \quad \tilde{\boldsymbol{\Delta}}_{\boldsymbol{R}} = \eta^{(a)}(i\omega_n)\boldsymbol{\Delta_R}. \tag{40}$$

To calculate the inverse of $\tilde{\mathcal{L}}_+$, we use an operator identity for $\omega_n > 0$,

$$\tilde{\mathcal{L}}_+^{-1} = \int_0^\infty dt \exp[-\tilde{\mathcal{L}}_+ t]. \tag{41}$$

Then we need to know how the exponential operator acts on the gap.

Since the Abrikosov lattice is a superposition of Φ_0 at different vortex cores, we introduce the lowering and raising operators,

$$a = \frac{\Lambda}{\sqrt{2}} \left(\pi_{x'} - i\pi_{y'}\right), \tag{42}$$

$$a^\dagger = \frac{\Lambda}{\sqrt{2}} \left(\pi_{x'} + i\pi_{y'}\right), \tag{43}$$

where $\pi'_{i'} = -i\nabla_{i'} - 2eA_{i'} = \xi_i\pi_i/\Lambda$ is the dynamical (gauge-invariant) momentum for the Cooper pair. The prime means that the isotropic coordinate is used according to the relation, (38). The operators satisfy the usual bosonic commutation relation $[a, a^\dagger] = 1$. Since $a\Delta_{\boldsymbol{R}} = 0$, the Abrikosov lattice (37) can be regarded as the vacuum state $|0\rangle$ in the language of the operator formalism. Applying $(a^\dagger)^m$ to the vacuum state we obtain the excited states,

$$|m\rangle = \sum_n C_n e^{-ip_n y'} \Phi_m[(x' - \Lambda^2 p_n)/\Lambda], \tag{44}$$

which corresponds to the m-th excited oscillator states centered on each vortices. Similarly, we introduce the raising and lowering operators as $b^\dagger = (a^\dagger)^*$ and $b = (a)^*$ acting on the conjugate vacuum state, $\Delta^*_{\boldsymbol{R}} = \langle 0|$, since $(a)^*\Delta^*_{\boldsymbol{R}} = 0$ holds. Owing to the factor $e^{ip_n y'}$, a spatial average ensures an orthogonal relation for different excited states as

$$\overline{\langle m|m'\rangle} = |\Delta|^2 \delta_{m,m'}, \tag{45}$$

since only functions Φ_m and $\Phi_{m'}$ centered on the same vortex contribute to the integral. Here we have defined the averaged gap magnitude as $|\Delta|^2 = \overline{|\Delta_{\boldsymbol{R}}|^2}$. Without loss of generality, we can take Δ real, because Δ appears only in the form $|\Delta|^2$. Note that the orthogonality does not hold without the spatial average.

In the operator formalism denoting

$$\boldsymbol{v}(\hat{\boldsymbol{k}}) = \left[v_\perp(\hat{\boldsymbol{k}}) \cos\phi_v, v_\perp(\hat{\boldsymbol{k}}) \sin\phi_v, v_\parallel(\hat{\boldsymbol{k}})\right], \tag{46}$$

we write

$$\tilde{\mathcal{L}}_+ = \tilde{\omega}_n + \frac{i}{\sqrt{2}} \left(\frac{\tilde{v}_\perp(\hat{\boldsymbol{k}})}{2\Lambda}\right) \left(e^{-i\phi'} a^\dagger + e^{i\phi'} a\right), \tag{47}$$

where we have defined

$$\begin{aligned} \tilde{v}_\perp(\hat{\boldsymbol{k}}) &= v_\perp(\hat{\boldsymbol{k}}) \sqrt{\chi^{-1/2} \cos^2\phi_v + \chi^{1/2} \sin^2\phi_v} \\ &= \sqrt{\chi^{-1/2} v_x^2 + \chi^{1/2} v_y^2}, \end{aligned} \tag{48}$$

$$\tan\phi' = \chi^{1/2} \tan\phi_v = \chi^{1/2} v_y/v_x, \tag{49}$$

with the anisotropy parameter, $\chi = \langle v_x^2\rangle/\langle v_y^2\rangle$. Note that $\tilde{v}_\perp(\hat{\boldsymbol{k}})$ is of the order of $[\langle v_x^2\rangle\langle v_y^2\rangle]^{1/4}$, and we obtain $\tilde{v}_\perp(\hat{\boldsymbol{k}}) = v_\perp(\hat{\boldsymbol{k}})$ and $\phi' = \phi_v$ in the absence of the anisotropy, i.e., $\chi = 1$. In introducing the anisotropy parameter, χ in eq. (48), we have related it to the anisotropy of the components of the Fermi velocity perpendicular to the field. The physical meaning of χ is the square of the ratio of coherence lengths, which takes into account the average contribution of the anisotropic shape of the vortex cores. Therefore, minimizing the free energy with respect to this parameter optimizes the shape of the vortex cores, and gives better results[57, 58, 59], especially in a comparison of different directions of the fields

Using Taylor expansions in $a^\dagger$ and a, we obtain

$$\tilde{\mathcal{L}}_+^{-1}\Delta_{\boldsymbol{R}} = \sqrt{\pi}\left(\frac{2\Lambda}{\tilde{v}_\perp(\hat{\boldsymbol{k}})}\right)\sum_{m=0}^{\infty} e^{-im\phi'} \times \frac{1}{\sqrt{m!}}\left(-\frac{1}{\sqrt{2}}\right)^m W^{(m)}(i\tilde{u}_n)|m\rangle, \tag{50}$$

with $\tilde{u}_n(\hat{\boldsymbol{k}}; i\omega_n) = [2\Lambda/\tilde{v}_\perp(\hat{\boldsymbol{k}})]\tilde{\omega}_n$, where $W(z) = e^{-z^2}\mathrm{erfc}(-iz)$ is the Faddeeva (normalized error) function and $W^{(m)}(z)$ denotes its m-th derivative. The properties of the function is summarized in the Appendix B. The conjugate version $\tilde{\mathcal{L}}_-^{-1}\Delta_{\boldsymbol{R}}^*$ is obtained by the replacements $\phi' \to -\phi'$ and $|m\rangle \to \langle m|$ in eq. (50). Then we obtain

$$\overline{\boldsymbol{f}_{\boldsymbol{R}}^\dagger \cdot \boldsymbol{f}_{\boldsymbol{R}}} = \frac{\sqrt{\pi}}{i}\left(\frac{2\Lambda}{\tilde{v}_\perp(\hat{\boldsymbol{k}})}\right)^2 \eta^{(a)\dagger}\eta^{(a)}|\boldsymbol{\Delta}(\hat{\boldsymbol{k}})|^2 W^{(1)}(i\tilde{u}_n)g^2, \tag{51}$$

where we have used the formula,

$$\sum_{m=0}^{\infty}\frac{1}{m!}\left(\frac{1}{2}\right)^m\left[W^{(m)}(i\tilde{u}_n)\right]^2 = \frac{1}{i\sqrt{\pi}}W^{(1)}(i\tilde{u}_n), \tag{52}$$

and we have abbreviated $\boldsymbol{\Delta}\varphi_\gamma(\hat{\boldsymbol{k}})$ to $\boldsymbol{\Delta}(\hat{\boldsymbol{k}})$.

Using the above result together with the normalization, $g^2 + \overline{\boldsymbol{f}_{\boldsymbol{R}}^\dagger \cdot \boldsymbol{f}_{\boldsymbol{R}}} = 1$, we finally obtain the normal propagator,

$$g = \left[1 + \frac{\sqrt{\pi}}{i}\left(\frac{2\Lambda}{\tilde{v}_\perp(\hat{\boldsymbol{k}})}\right)^2 \eta^{(a)\dagger}\eta^{(a)}|\boldsymbol{\Delta}(\hat{\boldsymbol{k}})|^2 W^{(1)}(i\tilde{u}_n)\right]^{-\frac{1}{2}}. \tag{53}$$

The quantity $I = \overline{I_{\boldsymbol{R}}}$ is also obtained as

$$I = \sqrt{\pi}\left(\frac{2\Lambda}{\tilde{v}_\perp(\hat{\boldsymbol{k}})}\right)|\boldsymbol{\Delta}(\hat{\boldsymbol{k}})|^2 W(i\tilde{u}_n)g\left[\eta^{(a)} + \eta^{(a)\dagger} - \frac{2g}{1+g}\eta^{(a)\dagger}\eta^{(a)}\right] + 2\omega_n(\eta^{(n)} - 1)(1-g). \tag{54}$$

In the clean limit, neglecting the impurity scattering rate, $\nu = 1/2\tau_0 T_c \ll 1$, we have

$$I = \frac{2g}{1+g}\sqrt{\pi}\left(\frac{2\Lambda}{\tilde{v}_\perp(\hat{\boldsymbol{k}})}\right)|\boldsymbol{\Delta}(\hat{\boldsymbol{k}})|^2 W(iu_n). \tag{55}$$

In the absence of B using $W(z) \sim i/\sqrt{\pi}z$ and $W^{(1)}(z) \sim -i/\sqrt{\pi}z^2$ for $|z| \gg 1$, we recover the uniform solution of eqs. (27) and (28). On the other hand, we get $g = 1$ and $I = I' = 0$ in the normal state, $\boldsymbol{\Delta} = 0$.

3.2 Self-consistent Equations for Renormalization Factors, $\eta^{(n)}(i\omega_n)$ and $\eta^{(a)}(i\omega_n)$

To complete the system of equations, we need self-consistencies for $\eta^{(n)}(i\omega_n)$ and $\eta^{(a)}(i\omega_n)$. Since it is impossible to obtain general analytic solutions, we assume that the structure of the impurity self-energies is similar to that in the absence of the magnetic field. With this assumption, we neglect the renormalization of the gap, i.e. $\langle f_{\boldsymbol{R}}\rangle \sim 0$, in the case of non-$s$-wave pairings, because the angular average with the factor $\varphi_\gamma(\hat{\boldsymbol{k}})$ gives small contribution to the self-energy (the angular average exactly vanishes in the absence of magnetic fields). We then obtain

$$\eta^{(n)} = 1 + \frac{1}{2\omega_n\tau_0}\frac{\langle g\rangle}{\cos^2\delta + \langle g\rangle^2\sin^2\delta}, \quad \eta^{(a)} = 1. \tag{56}$$

On the other hand, we consider only the Born limit $\delta = 0$ in the case of the isotropic s-wave singlet $\varphi_\gamma(\hat{\boldsymbol{k}}) = 1$. Retaining the term proportional to $\Delta_{\boldsymbol{R}}$ ($m = 0$ component) in eq. (50), namely,

$$f_{\boldsymbol{R}} \sim g\sqrt{\pi}\left(\frac{2\Lambda}{\tilde{v}_\perp(\hat{\boldsymbol{k}})}\right) W(i\tilde{u}_n)\Delta_{\boldsymbol{R}}, \tag{57}$$

(this is exact for an isotropic Fermi surface), we get

$$\begin{aligned} \eta^{(n)} &= 1 + \frac{1}{2\omega_n\tau_0}\langle g\rangle, \\ \eta^{(a)} &= \left[1 - \frac{\sqrt{\pi}}{2\tau_0}\left\langle g\left(\frac{2\Lambda}{\tilde{v}_\perp(\hat{\boldsymbol{k}})}\right) W(i\tilde{u}_n)\right\rangle\right]^{-1}. \end{aligned} \tag{58}$$

Note that the Anderson theorem ($\eta^{(n)}/\eta^{(a)} = 1$) is restored in the case of $B = 0$. Thus in all cases in the limit $B = 0$, we recover all the results of the standard theory for dirty superconductors. In the clean limit we have $\eta^{(n)}(i\omega_n) = \eta^{(a)}(i\omega_n) = 1$.

3.3 Thermodynamic Quantities

With the approximate analytic solutions obtained in the previous subsection, the free energy (32) can be regarded as a function of $\boldsymbol{\Delta}$ (can be taken as real) and B for given temperature T and the external field H. To obtain an equilibrium state, we minimize the free energy (32) with respect to $(B, \boldsymbol{\Delta})$. In searching the minimum point, we need to compute the free energy for given $(B, \boldsymbol{\Delta})$. To do this we first solve the self-consistent equations for the impurity renormalization $\eta^{(n)}(i\omega_n)$ and $\eta^{(a)}(i\omega_n)$ at positive ω_n's, i.e. eqs. (53) and (56) for non-s-wave pairings, or eqs. (53) and (58) for s-wave singlet. This can easily be done by an iteration starting from the clean limit, $\eta^{(n)} = \eta^{(a)} = 1$. Once we determine $\eta^{(n)}$ and $\eta^{(a)}$, we can compute the free energy (32) via eqs. (53) and (54). Note that all the numerical calculations can be done with real quantities in the Matsubara formalism, since $i^n W^{(n)}(ix)$ is real for real x. At $T = 0$ we simply replace $2\pi T\sum_{n=0}^{n_c} F(i\omega_n)$ with $\int_0^{2\pi n_c} d\omega F(i\omega)$, and

$$\ln\left(\frac{T}{T_c}\right) + 2\pi T\sum_{n=0}^{n_c}\frac{1}{\omega_n} = \ln(4e^\gamma n_c). \tag{59}$$

Once we obtain the equilibrium values of B and $\boldsymbol{\Delta}$, we can compute various thermodynamic quantities as the spatial average. The magnetization is given by

$$-4\pi M(T,H) = H - B(T,H). \tag{60}$$

Viewing the one-body nature of the quasiparticles in the present formalism, we can write down the entropy,

$$S(T,H) = -4\int_0^\infty d\omega N(\omega;T,H)\left[f\left(\frac{\omega}{T}\right)\ln f\left(\frac{\omega}{T}\right) + \left(1 - f\left(\frac{\omega}{T}\right)\right)\ln\left(1 - f\left(\frac{\omega}{T}\right)\right)\right], \tag{61}$$

with $f(x) = (1+e^x)^{-1}$ via the DOS in the superconducting state,

$$\frac{N(\omega;T,H)}{N_0} = \mathrm{Re}\left\langle g(\hat{\boldsymbol{k}}; i\omega_n \to \omega + i\delta)\right\rangle, \tag{62}$$

where we have done the analytic continuation from $i\omega_n$ in the upper half plane to $\omega + i\delta$, δ being an infinitesimal positive number. Then the specific heat is calculated numerically by $C(T,H)/T = \partial S/\partial T$.

It is useful to discuss the region $B \sim H \sim H_{c2}$ in the clean limit, where $\boldsymbol{\Delta}$ is suppressed. Expanding the free energy in $\boldsymbol{\Delta}$, we obtain the free energy under magnetic fields,

$$\Omega_{\mathrm{SN}} \sim \frac{(B-H)^2}{8\pi} + N_0\alpha(T,B)|\boldsymbol{\Delta}|^2 + \frac{1}{2}\frac{N_0}{T_c^2}\beta(T,B)\left(|\boldsymbol{\Delta}|^2\right)^2, \tag{63}$$

where

$$\alpha(T,B) = \left\langle\left[\ln\left(\frac{T}{T_c}\right) + 2\pi T\sum_{n=0}^{\infty}\left\{\frac{1}{\omega_n} - \sqrt{\pi}\left(\frac{2\Lambda}{\tilde{v}_\perp(\hat{\boldsymbol{k}})}\right)W(iu_n)\right\}\right]|\varphi_\gamma(\hat{\boldsymbol{k}})|^2\right\rangle, \tag{64}$$

and

$$\beta(T,B) = -i\pi^2 T T_c^2 \sum_{n=0}^{\infty}\left\langle\left(\frac{2\Lambda}{\tilde{v}_\perp(\hat{\boldsymbol{k}})}\right)^3 \times W(iu_n)W^{(1)}(iu_n)|\varphi_\gamma(\hat{\boldsymbol{k}})|^4\right\rangle. \tag{65}$$

The temperature dependence of the upper critical field is determined by

$$\alpha(T, H_{c2}) = 0. \tag{66}$$

In the limit $H_{c2} \to 0$ we recover the BCS results, i.e. $\alpha = \ln(T/T_c)$ and $\beta = [7\zeta(3)/8\pi^2]\langle|\varphi_\gamma(\hat{\boldsymbol{k}})|^4\rangle$, where $\zeta(n)$ is the Riemann zeta function[1]. The field dependence of the magnetization and the gap near H_{c2} are given as

$$-4\pi M = 4\pi\alpha'_{c2}(T)N_0|\boldsymbol{\Delta}|^2 = \left(\frac{\beta_{c2}(T)}{4\pi\alpha'^2_{c2}(T)N_0T_c^2} - 1\right)^{-1}(H_{c2} - H), \tag{67}$$

where $\alpha'_{c2}(T) = \partial\alpha(T,B)/\partial B|_{B=H_{c2}}$ and $\beta_{c2}(T) = \beta(T, H_{c2})$. At $T = 0$ we can perform the summation over ω_n in (64) and (65). Then we obtain

$$\alpha(0,B) = \frac{1}{2}\ln\left(\frac{B}{H_{c2}}\right), \tag{68}$$

with the upper critical field at $T = 0$,

$$H_{c2} = \frac{2}{|e|}\left(\frac{\pi T_c}{v}\right)^2 \exp\left[-\left\langle |\varphi_\gamma(\hat{\boldsymbol{k}})|^2 \ln\left(\frac{\tilde{v}_\perp(\hat{\boldsymbol{k}})}{v}\right)^2\right\rangle - \gamma\right], \tag{69}$$

where v without $(\hat{\boldsymbol{k}})$ denotes the angular average of $|\boldsymbol{v}(\hat{\boldsymbol{k}})|$ and $\gamma \doteqdot 0.577216$ is the Euler's constant, and

$$\beta(0, B) = \frac{\pi}{2|e|B}\left(\frac{T_c}{v}\right)^2 \left\langle \left(\frac{v}{\tilde{v}_\perp(\hat{\mathbf{k}})}\right)^2 |\varphi_\gamma(\hat{\mathbf{k}})|^4 \right\rangle. \tag{70}$$

In the case of strong anisotropy, $\chi \ll 1$ using $\tilde{v}_\perp / v_\perp \sim \chi^{1/4}$, we obtain

$$\frac{H_{c2}(T=0;\chi)}{H_{c2}(T=0;1)} \sim \chi^{-1/2} = \sqrt{\frac{\langle v_y^2 \rangle}{\langle v_x^2 \rangle}}. \tag{71}$$

Before closing this subsection, we determine the critical value of the Ginzburg-Landau parameter, κ to separate the type I and II superconductors. Since the quantity, $H_{c2}/T_c\sqrt{N_0}$, is dimensionless and proportional to the ratio between the penetration depth and the coherence length, we define it as the GL parameter. At $T = 0$ in the Meissner state ($B = 0$), the free energy is written as

$$\frac{\Omega_{\mathrm{SN}}}{N_0 T_c^2} = \frac{\kappa^2}{8\pi}\left(\frac{H}{H_{c2}}\right)^2 + \frac{1}{2}\left(\frac{\Delta}{T_c}\right)^2 \left[\ln\left(\frac{\Delta}{\Delta_0}\right)^2 - 1\right], \tag{72}$$

where the gap magnitude at $T = B = 0$[1] is given by

$$\frac{\Delta_0}{T_c} = \frac{\pi}{e^\gamma} \exp\left[-\frac{1}{2}\langle |\varphi(\boldsymbol{k})|^2 \ln |\varphi(\boldsymbol{k})|^2 \rangle\right]. \tag{73}$$

Thus, the minimum of the free energy is obtained by $\Delta = \Delta_0$ as it should be. The thermodynamic critical field is then obtained by equating $\Omega_{\mathrm{SN}}(H_c) = 0$:

$$\frac{H_c}{H_{c2}} = \frac{\sqrt{4\pi}}{\kappa}\frac{\Delta_0}{T_c}. \tag{74}$$

At the critical value of κ we have $H_{c1} = H_c = H_{c2}$. It gives the critical value of the GL parameter as

$$\kappa_c = \sqrt{4\pi}\frac{\Delta_0}{T_c}. \tag{75}$$

3.4 Extension to Multiband Superconductors

In a system having plural Fermi surfaces, each band could have different symmetry and/or magnitude of gaps due to symmetry reason [60, 61, 62]. Since those Fermi surfaces have different shape in general, the Cooper pair may be formed among the same band in order to reduce the energy associated with the center-of-mass motion. In this case the MF Hamiltonian (1) is diagonal in the band indices ℓ, and the quasiclassical propagators are also diagonal. Moreover, we do not take into account the interband impurity scattering. It is known that the mixing effect between different bands disappears both in the weak and the strong-coupling limit of the interband impurity scattering, hence the multiband situation is restored [63, 7]. Therefore, the coupling between bands only appears in the gap equation as

$$\sum_{\ell'} \left(V^{-1}\right)_{\ell\ell'} \boldsymbol{\Delta}_{\ell'\boldsymbol{R}} = \pi T N_{0\ell} \sum_{n} \left\langle \varphi^{*}_{\ell\gamma}(\hat{\boldsymbol{k}}) \boldsymbol{f}_{\ell\boldsymbol{R}}(\hat{\boldsymbol{k}}; i\omega_n) \right\rangle, \tag{76}$$

with $\boldsymbol{\Delta}_{\ell\boldsymbol{R}}(\hat{\boldsymbol{k}}) = \boldsymbol{\Delta}_{\ell\boldsymbol{R}}\varphi_{\ell\gamma}(\hat{\boldsymbol{k}})$, which corresponds to eq. (31). Here we have used

$$V_{\ell\ell'}(\hat{\boldsymbol{k}}, \hat{\boldsymbol{k}}') = -\left(V\right)_{\ell\ell'} \sum_{\gamma} \varphi_{\ell\gamma}(\hat{\boldsymbol{k}}) \varphi^{*}_{\ell'\gamma}(\hat{\boldsymbol{k}}'), \tag{77}$$

where the coupling constant V is a matrix in the band space.

The free energy scaled by $N_0 T_c^2$, $N_0 = \sum_\ell N_{0\ell}$ being the total DOS in the normal state, is given by

$$\frac{\Omega_{\rm SN}}{N_0 T_c^2} = \frac{\kappa^2}{8\pi}(b-h)^2 + \sum_{\ell\ell'} \alpha^{(0)}_{\ell\ell'} \boldsymbol{d}^{*}_{\ell} \cdot \boldsymbol{d}_{\ell'} + \sum_{\ell} n_\ell \left[|\boldsymbol{d}_\ell|^2 \ln t + 2\pi t \sum_{n=0}^{\infty} \left(\frac{|\boldsymbol{d}_\ell|^2}{o_n} - \langle i_\ell \rangle - i'_\ell \right) \right], \tag{78}$$

with

$$\alpha^{(0)}_{\ell\ell'} = (\lambda^{-1})_{\ell\ell'} - n_\ell A_c \delta_{\ell\ell'}, \tag{79}$$

where $\lambda = N_0 V$, $n_\ell = N_{0\ell}/N_0$, $(b, h) = (B, H)/H_{c2}(T = 0)$ and $(t, \boldsymbol{d}_\ell, o_n, i_\ell, i'_\ell) = (T, \boldsymbol{\Delta}_\ell, \omega_n, I_\ell, I'_\ell)/T_c(H = 0)$. Since all the argument in the previous section holds for each band ℓ, we have added the subscript ℓ to the relevant quantities. Note that the angular average should be carried out over the corresponding Fermi surface. Owing to independent character of the quasiparticles in each bands, the entropy and the specific heat can be obtained by adding each contributions with the weight n_ℓ (the primary order parameter is specified by $\ell = 1$). In the presence of the anisotropy, each χ_ℓ should independently be determined[59]. The use of the Abrikosov lattice with different χ_ℓ's in different bands results in unphysical situation since the structure of penetrating flux differs in each bands. Nevertheless, the independent optimization of χ_ℓ respects the fact that the vortex cores have different shapes in each bands, which is confirmed by both theoretically[64] and experimentally[65] in the case of the typical two-gap superconductor, $\mathrm{MgB_2}$. We should recall again that the present formalism has a meaning only as the spatial average.

The factor $A_c = \ln(2e^\gamma \omega_c/\pi T_c)$ is the smallest eigenvalue determined by

$$\det \alpha^{(0)}_{\ell\ell'} = 0. \tag{80}$$

It is useful to introduce a dimensionless critical field $b_{c2} = |e|(v_1/T_c)^2 H_{c2}/2$, which will be determined such that the gap vanishes at $b = 1$ and $t = 0$.

Using the argument similar to the single-band case, we obtain the free energy near the upper critical fields in the clean limit,

$$\frac{\Omega_{\rm SN}}{N_0 T_c^2} \sim \frac{\kappa^2}{8\pi}(b-h)^2 + \sum_{\ell\ell'} \left(\alpha^{(0)}_{\ell\ell'} + \alpha_\ell(t,b)\delta_{\ell\ell'} \right) \boldsymbol{d}^{*}_{\ell} \cdot \boldsymbol{d}_{\ell'} + \frac{1}{2} \sum_{\ell} \beta_\ell(t,b) \left(|\boldsymbol{d}_\ell|^2 \right)^2, \tag{81}$$

where

$$\alpha_\ell(t,b) = n_\ell\Bigg\langle\Bigg[\ln t + 2\pi t\sum_{n=0}^{\infty}\Bigg\{\frac{1}{o_n} - \frac{\sqrt{\pi}}{\sqrt{bb_{c2}}}\left(\frac{v_1}{\tilde{v}_{\ell\perp}(\hat{\boldsymbol{k}})}\right)\times W(iu_{\ell n})\Bigg\}\Bigg]|\varphi_{\ell\gamma}(\hat{\boldsymbol{k}})|^2\Bigg\rangle, \quad (82)$$

and

$$\beta_\ell(t,b) = -n_\ell i\pi^2 t\sum_{n=0}^{\infty}\left\langle(bb_{c2})^{-3/2}\left(\frac{v_1}{\tilde{v}_{\ell\perp}(\hat{\boldsymbol{k}})}\right)^3\times W(iu_{\ell n})W^{(1)}(iu_{\ell n})|\varphi_{\ell\gamma}(\hat{\boldsymbol{k}})|^4\right\rangle. \quad (83)$$

Note that only the quadratic term contains the coupling between different bands, which corresponds to the internal Josephson coupling. We obtain the temperature dependence of the upper critical field, $b(t)$ by solving the equation,

$$\det\left(\alpha_{\ell\ell'}^{(0)} + \alpha_\ell(t,b)\delta_{\ell\ell'}\right) = 0. \quad (84)$$

In particular, at $T = 0$ we have

$$\alpha_\ell(0,b) = n_\ell\left\langle\ln\left[\frac{e^{\gamma/2}}{\pi}\sqrt{bb_{c2}}\left(\frac{\tilde{v}_{\ell\perp}(\hat{\boldsymbol{k}})}{v_1}\right)\right]|\varphi_{\ell\gamma}(\hat{\boldsymbol{k}})|^2\right\rangle, \quad (85)$$

and

$$\beta_\ell(0,b) = n_\ell\frac{\pi}{4bb_{c2}}\left\langle\left(\frac{v_1}{\tilde{v}_{\ell\perp}(\hat{\boldsymbol{k}})}\right)^2|\varphi_{\ell\gamma}(\hat{\boldsymbol{k}})|^4\right\rangle. \quad (86)$$

It is also noted that eq. (78) is valid for multi-component gap functions[58]. Namely, we can consider $\boldsymbol{\Delta}(\boldsymbol{k}) = \Delta_{\boldsymbol{R}}\sum_\gamma \eta_\gamma\varphi_\gamma(\boldsymbol{k})$ with the normalization, $\sum_\gamma\langle|\varphi_\gamma(\boldsymbol{k})|^2\rangle = \sum_\gamma|\eta_\gamma|^2 = 1$.

4 Comparison with Numerical Results

Now we demonstrate the application of our theory to typical pairings, i.e. (a) the single-band isotropic s-wave, (b) the single-band $d_{x^2-y^2}$-wave and (c) the two-band s-waves. For simplicity, we consider strongly type-II superconductors ($\kappa \gg 1$) in the clean limit with two-dimensional cylindrical Fermi surfaces. The magnetic field is applied along z axis. We also use the same Fermi velocity for two bands. Thus we have $B = H$, $\chi = 1$, $v_1/\tilde{v}_{\ell\perp}(\hat{\boldsymbol{k}}) = 1$, $\langle\cdots\rangle = \int_0^{2\pi}\frac{d\phi}{2\pi}\cdots$, and only the gap(s) are to be determined.

Let us first show results for the single-band isotropic s-wave, $\varphi(\hat{\boldsymbol{k}}) = 1$. Figure 1 shows the averaged thermodynamic quantities in the mixed state, (a) the field dependence of the gap magnitude Δ scaled by $\Delta_0 = \Delta(h = t = 0)$ at different temperatures, (b) the field dependence of the DOS at $t = 0$, (c) the field dependence of the zero-energy (ZE) DOS at $t = 0.1$, and (d) the temperature dependence of the specific heat normalized to that of the normal state, C_N. In Fig. 1(a) the gap at the low temperature first increases as the field increases. This is an drawback of the present approximation, which is not valid at very low fields at low temperature. The thin line represents an empirical formula,

$\Delta(t,h) = \Delta(t,0)\sqrt{1-h}$, which well describes the field dependence of the gap for $t > 0.5$. In Fig. 1(b) the peak position of the DOS moves upward in energy with the external field. In Fig. 1(c) the open circles are taken from the results of the full numerical solution of the quasiclassical equations done by Miranović *et al.* [66, 67]. It indicates that the BPT approximation works very well for $h > 0.6$ at low temperatures. For lower fields the BPT approximation overestimates contribution from vortex cores since the vortex core size becomes unphysically large in the Abrikosov lattice model, (37) (see numerical results [66]).

Next we move to the case of $d_{x^2-y^2}$-wave, $\varphi(\hat{\boldsymbol{k}}) = \sqrt{2}\cos(2\phi)$. Figure 2 shows results for $d_{x^2-y^2}$-wave in the plot similar to the previous case. The same empirical formula works fine for $t > 0.5$. A tendency of the peak shift in the DOS is the similar to s-wave, but an amount of the shift is smaller than that of s-wave. Although the BPT approximation overestimates the vortex core contribution in comparison with full numerical results [67, 68], it shows better agreement than the case of s-wave.

Figure 3 shows results for two-band s-waves. We use the same density of states for both bands, $n_1 = n_2 = 0.5$ for simplicity. Here the coupling matrix is given by

$$\lambda_{\ell\ell'} = \begin{pmatrix} \lambda_1 & \lambda \\ \lambda & \lambda_2 \end{pmatrix}, \tag{87}$$

with $\lambda_1 = 0.25$, $\lambda_2 = 0.08\lambda_1$ and $\lambda = 0.26\lambda_1$. In Fig. 3(c) the ZEDOS's are compared with numerical calculation of the Bogoliubov de Gennes framework at $T = 0$ [13]. The field dependence in the passive band agrees very well with the full numerical result since the vortex core size is sufficiently large due to the smallness of the gap with large coherence length. The discrepancy mainly comes from the primary band but its contribution accounts for n_1 of the total ZEDOS. Therefore, the discrepancy in total becomes less remarkable. It should be emphasized that behaviors of two-band systems would be changed sensitively by a slight change of material parameters and a combination of two gaps. Therefore in discussing experimental data it is necessary to take into account precise material parameters in accordance with band-structure calculations [8, 10, 11].

Finally, we demonstrate the magnetization curves for several κ's in the case of s-wave in Fig. 4. According to eq. (75), the critical GL parameter, $\kappa_c = 2\pi^{3/2}e^{-\gamma} \doteqdot 6.25278$ for s-wave. The thin lines denote results obtained by means of the GL theory [69]. For larger κ, the BPT approximation shows better agreement with the GL results. In the BPT approximation, the transition at the lower critical field, H_{c1} becomes first order for $\kappa/\kappa_c < 1.9$, i.e., the free energy has two minima in (B, Δ) space above H_{c1}. In this case, the hysteresis loop is represented by the arrow. For small κ nonlocal effect becomes important. Although the BPT approximation properly takes into account nonlocal effect beyond the GL theory, the treatment of averaged magnetic field itself becomes poor. To conclude the validity of the BPT approximation for small κ, we should await results of full numerical calculation.

5 Field Angle Dependence

For microscopic understanding of the unconventional superconductors, a determination of their gap symmetry is highly desired. Although power-law behaviors in thermodynamic

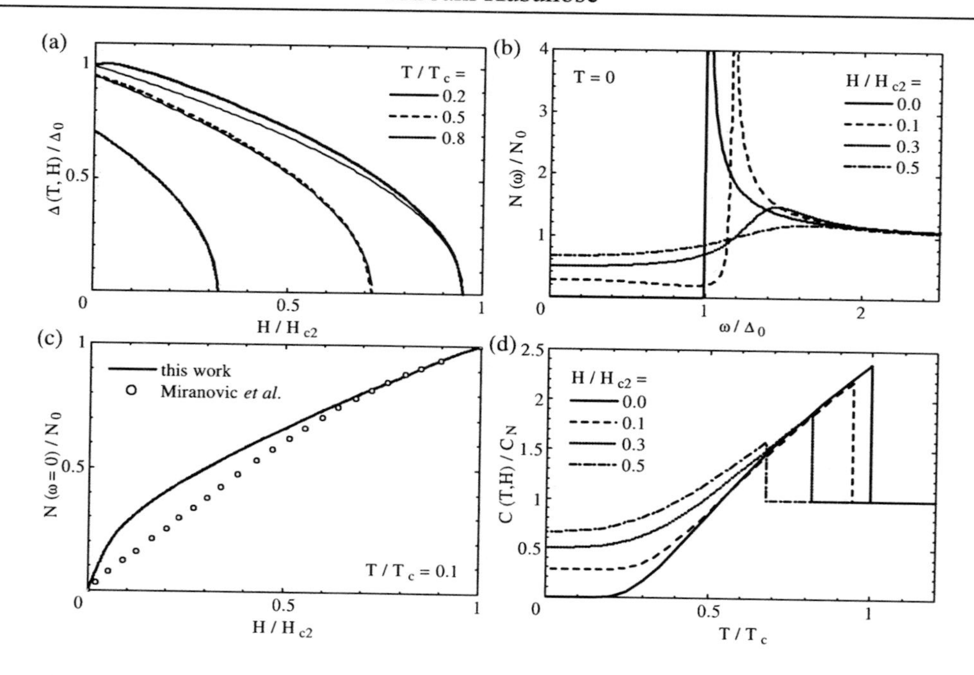

Figure 1: The averaged thermodynamic quantities for the s-wave: (a) the gap magnitude scaled by $\Delta_0 = \Delta(T = H = 0)$. The thin line represents the empirical formula $\Delta(T, H) = \Delta(T, 0)\sqrt{1 - H/H_{c2}}$, (b) the density of states at $T = 0$, (c) the field dependence of the zero-energy DOS compared with full numerical results [66, 67], (d) the specific heat.

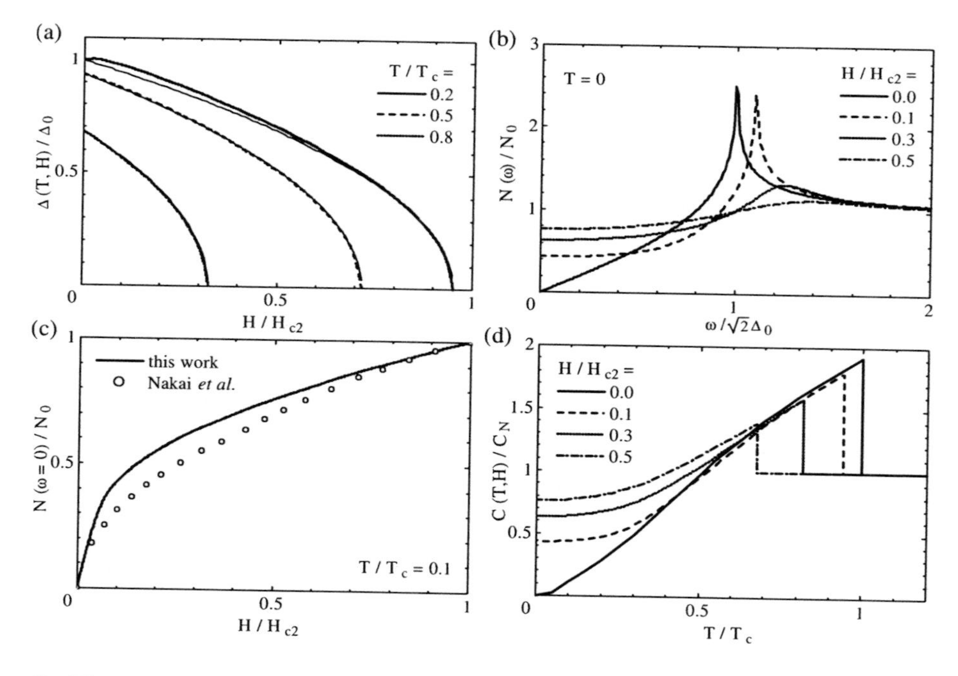

Figure 2: The averaged thermodynamic quantities for the $d_{x^2-y^2}$-wave: (a) the gap magnitude. Thin line is the same empirical formula in Fig. 1, (b) the DOS at $T = 0$, (c) the ZEDOS and full numerical results [67], (d) the specific heat.

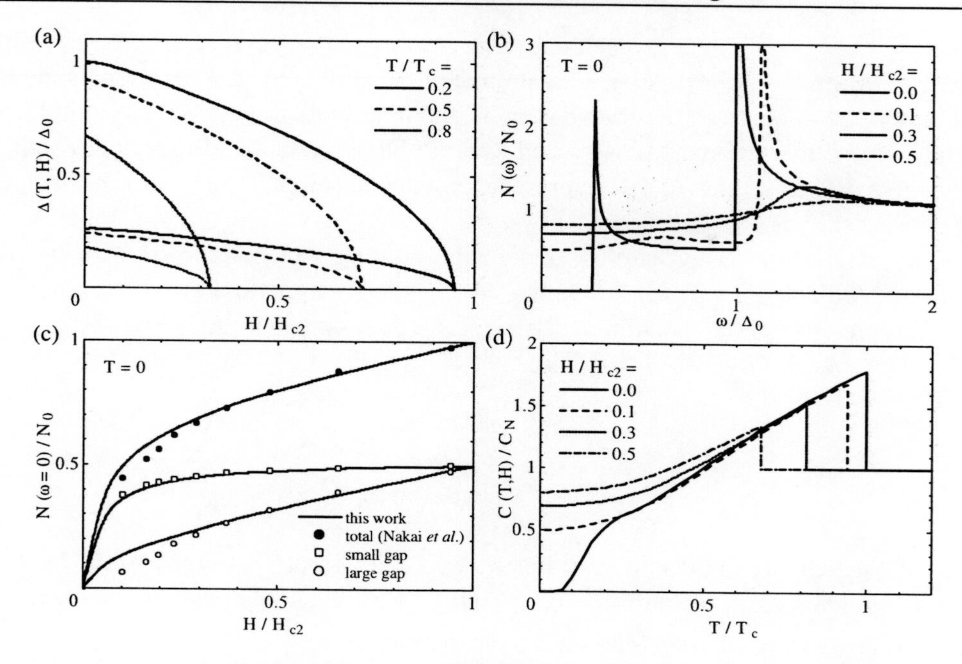

Figure 3: The averaged thermodynamic quantities for the two-band s-wave: (a) the gap magnitude scaled by the primary gap $\Delta_0 = \Delta_1(T = H = 0)$, (b) the DOS at $T = 0$, (c) the ZEDOS and numerical results [13], (d) the total specific heat. The coupling matrix is given by $\hat{\lambda} = ((\lambda_1, \lambda), (\lambda, \lambda_2))$ with $\lambda_1 = 0.25$, $\lambda_2 = 0.08\lambda_1$ and $\lambda = 0.26\lambda_1$.

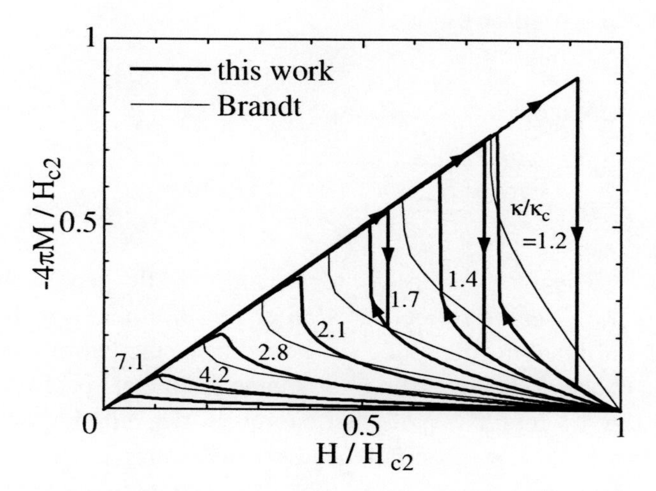

Figure 4: The magnetization curves for several GL parameters. The thin lines are taken from the GL result [69]. For $\kappa/\kappa_c < 1.9$ the lower transition becomes first order in the BPT approximation. The arrow indicates the hysteresis loop.

properties at low temperature give a hint of the structure of the gap function, it is not enough to draw definite conclusions. As was mentioned in the introduction, a powerful method has been developed to measure the modulation of gap amplitude directly by using field angle dependences of thermal conductivity and specific heat. Here we discuss the oscillatory behaviors in H_{c2} and C based on the present formalism. The influence of the anisotropy in the Fermi velocity is also discussed.

In order to discuss properties under the rotating magnetic field, we move to the new coordinate (x', y', z') in which the applied magnetic field is parallel to z' axis. Representing the rotation matrix $\hat{R}$ for the original magnetic field $\boldsymbol{H} = H(\sin\theta_h\cos\phi_h, \sin\theta_h\sin\phi_h, \cos\theta_h)$ as,

$$\hat{R} = \begin{pmatrix} \cos\theta_h\cos\phi_h & \cos\theta_h\sin\phi_h & -\sin\theta_h \\ -\sin\phi_h & \cos\phi_h & 0 \\ \sin\theta_h\cos\phi_h & \sin\theta_h\sin\phi_h & \cos\theta_h \end{pmatrix}, \tag{88}$$

we obtain the velocity in the new coordinate as $\boldsymbol{v}'_\ell(\hat{\boldsymbol{k}}) = \hat{R}\boldsymbol{v}_\ell(\hat{\boldsymbol{k}})$. In the case of $\boldsymbol{H} \parallel x$ for example, we have $v_{x'}^2 = v_z^2$ and $v_{y'}^2 = v_y^2$ with $\theta_h = \pi/2$ and $\phi_h = 0$ as they should be. Then, we can express the necessary quantities in terms of $\boldsymbol{v}'_\ell(\hat{\boldsymbol{k}})$ as

$$\chi_\ell = \langle v_{\ell x'}^2(\hat{\boldsymbol{k}})\rangle / \langle v_{\ell y'}^2(\hat{\boldsymbol{k}})\rangle, \tag{89}$$

$$\tilde{v}_{\ell\perp}(\hat{\boldsymbol{k}}) = \sqrt{\chi_\ell^{-1/2} v_{\ell x'}^2(\hat{\boldsymbol{k}}) + \chi_\ell^{1/2} v_{\ell y'}^2(\hat{\boldsymbol{k}})}. \tag{90}$$

For details of $\boldsymbol{v}_\ell(\hat{\boldsymbol{k}})$ we can refer to band structure calculations or simple tight-binding models.

5.1 Effect of Gap Anisotropy

Let us discuss the effect of the gap modulation[41, 42] and the Fermi velocity anisotropy[42], we take the case of Sr_2RuO_4[5] as an example. We consider the gap function,

$$\varphi(\hat{\boldsymbol{k}}) = \sqrt{\frac{\sin^2(\pi R\cos\phi) + \sin^2(\pi R\sin\phi)}{1 - J_0(2\pi R)}} \times \mathrm{sgn}(\sin\phi), \tag{91}$$

which is proposed by Miyake and Narikiyo for the model of the gap function in the primary γ band [70]. Here $J_0(x)$ is the Bessel function of zeroth order, ϕ is the azimuthal angle of $\hat{\boldsymbol{k}}$. Although the original meaning of R is the radius of the Fermi circle in unit of π/a, a being the lattice constant, we regard it as a phenomenological parameter to characterize the magnitude of the gap minima. Note that the results depend only on the magnitude of the gap, $|\varphi(\hat{\boldsymbol{k}})|$, so that there is no difference between isotropic s wave and the chiral p-wave for example. The angle dependence of the gap magnitude is shown in Fig. 5. $R = 1$ corresponds to the gap with line of zeros, while $R = 0$ is the isotropic full gap. For comparison, the dependence of f wave, $\varphi(\hat{\boldsymbol{k}}) = \sqrt{2}\sin(2\phi)e^{i\phi}$, is also shown in Fig. 5. It is also noted that only the relative field angle from the gap minima has a meaning, so that the qualitative results can be applied to other type of gap functions in two-dimensional systems.

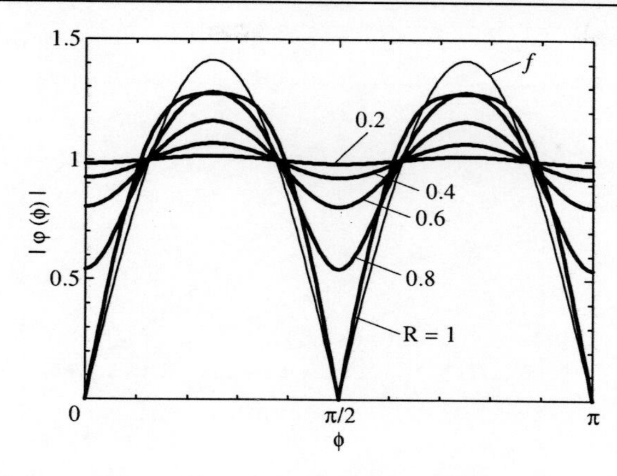

Figure 5: The angle dependence of the gap magnitude for various R. The thin line represents the case of f wave for comparison.

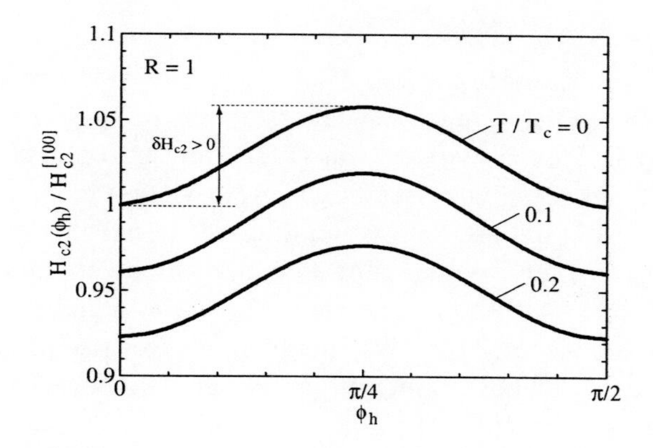

Figure 6: The in-plane field angle dependence of the upper critical field with line of zeros in [100] direction.

At $T = 0$ the in-plane H_{c2} anisotropy due to the modulation of the gap and/or the anisotropy of $\boldsymbol{v}$ is given by eq. (69). From this expression the minima of H_{c2} are obtained when the minima of the oscillation in the gap and in the perpendicular component of the velocity coincide. Thus, without the in-plane anisotropy in $\boldsymbol{v}$, the minima of H_{c2} are simply realized when the field is parallel to the direction of the gap minima.

Figure 6 shows the in-plane field angle dependence of H_{c2} for $R = 1$ at $T/T_c = 0$, 0.1 and 0.2. As we expect it shows minima for $\phi_h \parallel [100]$ ($= 0$ and $\pi/2$). The oscillatory amplitude in what follows is defined as $\delta A = A(\phi_h \parallel \text{antinode}) - A(\phi_h \parallel \text{node})$ where A is either H_{c2} or C. The temperature dependence of δH_{c2} scaled by $H_{c2}^{[100]}$ ([100] denotes the field direction.) is shown in Fig. 7. The amplitude of the H_{c2} oscillation is a monotonically decreasing function of T, and it becomes larger the smaller the size of the gap minimum is. The inset shows the temperature dependence of H_{c2} for $H \parallel [100]$.

The field dependence of δC at $T/T_c = 0.1$ is shown in Fig. 8. As the field decreases, the negative amplitude increases and changes sign at H^*. It is emphasized that even for the gap with line of zeros, $R = 1$, it reaches the maximum, and again decreases toward

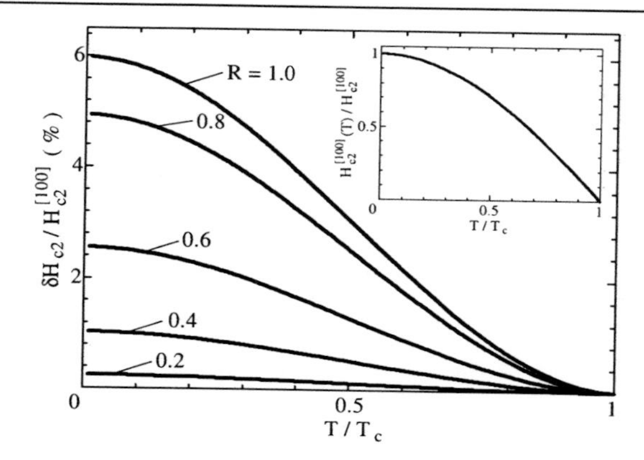

Figure 7: The temperature dependence of the oscillatory amplitude of H_{c2} with different size of the gap minimum, R. The inset shows the temperature dependence of H_{c2} for $H \parallel [100]$.

to zero. For larger modulation of the gap δC has stronger field dependence. The overall behavior at the lowest temperature can be understood as follows. At higher fields the in-plane H_{c2} anisotropy yields the oscillation in H/H_{c2} since H is fixed in the experimental situation. It leads to the amplitude oscillation in Δ, which predominates the oscillatory behavior in C. Therefore, the oscillatory amplitude of C has the opposite sign of δH_{c2}. At lower fields $H \lesssim H^*$, the Doppler shift argument can be applied [40]. The local QP with the momentum $\hat{\boldsymbol{k}}$ has the energy spectrum, $E_{\hat{\boldsymbol{k}}} + \boldsymbol{v}_{\rm s} \cdot \hat{\boldsymbol{k}}$ in the supercurrent flowing around vortices with the velocity $\boldsymbol{v}_{\rm s}$. This energy shift gives rise to a finite DOS near the gap minima if $|\hat{\boldsymbol{k}} \cdot \boldsymbol{v}_{\rm s}| > |\Delta(\hat{\boldsymbol{k}})|$. Since $\boldsymbol{v}_{\rm s} \perp \boldsymbol{H}$, the number of activated gap minima becomes largest in the case of H parallel to nodes yielding a maximum in C. In the limit of $H \to 0$, δC decreases again since the volume fraction of the locally activated QPs becomes small.

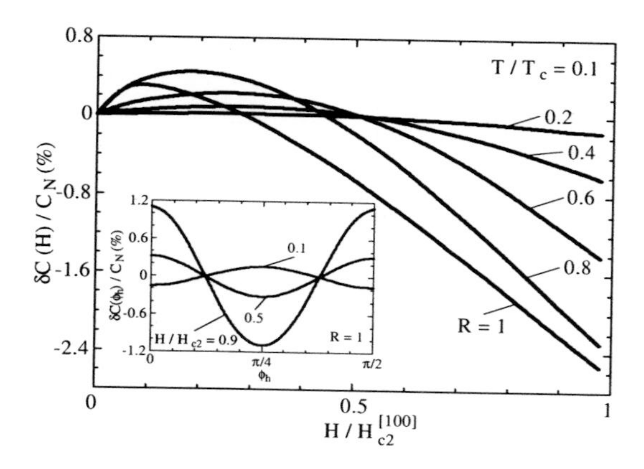

Figure 8: The field dependence of the oscillatory amplitude of the specific heat at $T/T_c = 0.1$. The inset shows the angle dependence of δC.

It should be noted that the Doppler shift argument is valid only at low temperature. At elevated temperature, positive contribution to δC from the activated QPs near the gap minima becomes less pronounced and the negative contribution from the H_{c2} anisotropy predominates. This is seen clearly by the field dependence of δC at $T/T_c = 0.1, 0.2$ and 0.3 as shown in Fig. 9. For $T/T_c \geq 0.2$ no sign changes occur in δC.

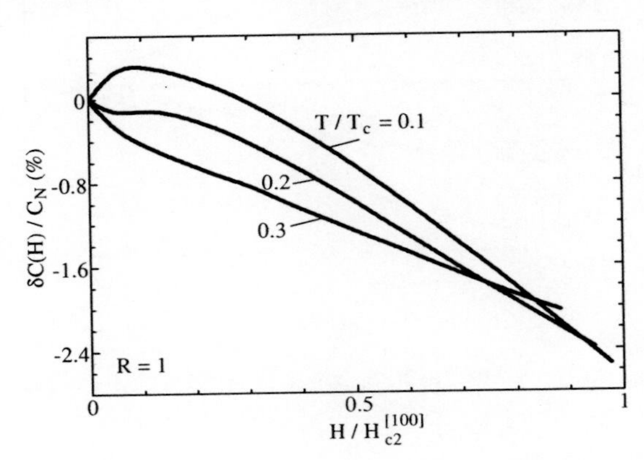

Figure 9: The oscillatory amplitude of specific heat at elevated temperature.

5.2 Effect of Fermi Velocity Anisotropy

Next we discuss the effect of the anisotropy of $\boldsymbol{v}$. For this purpose we consider the isotropic gap, $R = 0$. The essential feature of the anisotropy with 4-fold symmetry is described by

$$v_{x,y}(\hat{\boldsymbol{k}}) = \sqrt{\frac{2}{2+2a+a^2}} v_\perp \times \times \left[(1 + a\cos 2\phi)\cos\phi, (1 - a\cos 2\phi)\sin\phi\right]. \quad (92)$$

The anisotropy of $v_\perp(\phi) = \sqrt{v_x^2 + v_y^2}$ is shown in Fig. 10. The modulation of $\boldsymbol{v}$ along z direction is approximated roughly by $v_z(\hat{\boldsymbol{k}}) = v_z \mathrm{sgn}(k_z)$ for quasi two-dimensional Fermi surfaces. The two dimensionality is characterized by the ratio, $\chi = \langle v_z^2\rangle/\langle v_x^2\rangle = 2v_z^2/v_\perp^2$. The average of the Fermi velocity is defined as $v = \sqrt{\langle v_x^2 + v_y^2 + v_z^2\rangle} = \sqrt{v_\perp^2 + v_z^2}$.

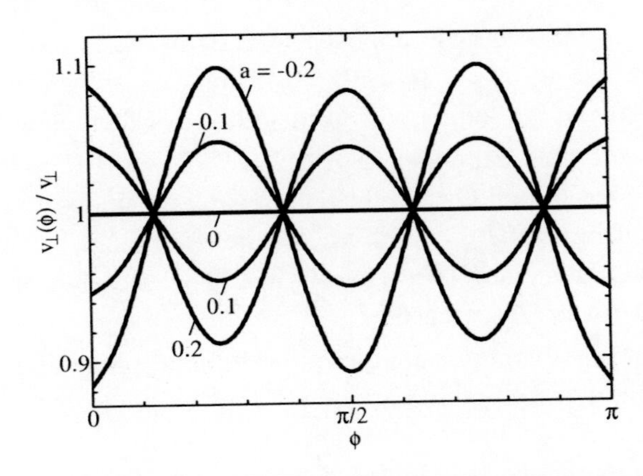

Figure 10: The in-plane anisotropy of the Fermi velocity

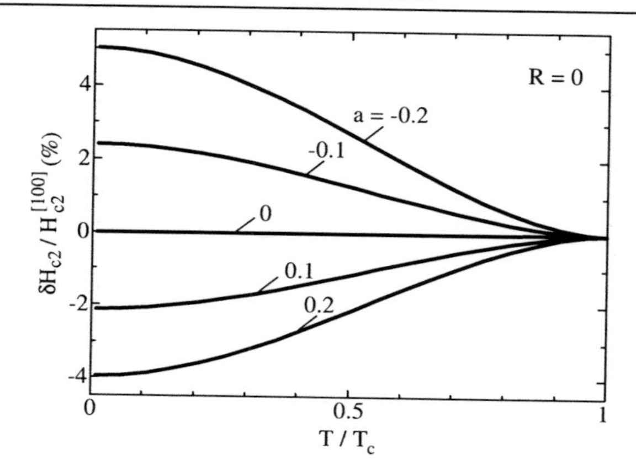

Figure 11: The temperature dependence of δH_{c2} for the isotropic full gap with the anisotropic Fermi velocity.

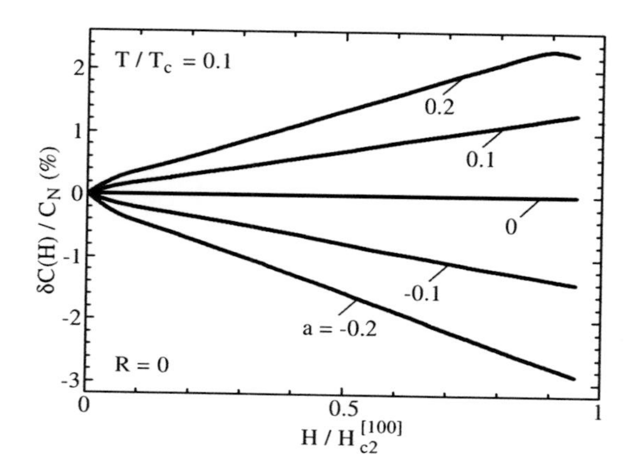

Figure 12: The field dependence of δC for the isotropic full gap with the anisotropic Fermi velocity.

The temperature dependence of δH_{c2} with the anisotropy is shown in Fig. 11. The effect of anisotropy in δH_{c2} has the opposite sign of a and is monotonically decreasing (increasing) function of T for $a < 0$ ($a > 0$). Note that the oscillatory amplitude due to the anisotropic velocity can be the same order of magnitude of that due to the gap modulation.

Figure 12 shows the field dependence of δC at $T/T_c = 0.1$. Since there is no contribution from the Doppler-shifted QPs in this case ($R = 0$), the oscillatory behaviors come purely from the anisotropy of H_{c2}, which affects the oscillatory behaviors over weak field region. Thus, the oscillatory amplitude of δC has the same sign of a and its magnitude decreases monotonically as H decreases.

It has been shown that the minimum direction of the in-plane H_{c2} anisotropy coincides with either the minimum of the gap modulation or the Fermi velocity. If both exists and has minima in different directions, the resultant amplitude becomes small due to the competing contributions from both effects. Due to the H_{c2} anisotropy the oscillatory behavior of C at high fields shows the opposite tendency of the H_{c2} anisotropy and its magnitude increases toward H_{c2}. The absolute magnitude of the H_{c2} anisotropy is a monotonically decreasing function of T, and hence, the effect of the Fermi velocity anisotropy is less pronounced in low temperatures and low fields. On the other hand, in these regions, the effect of the gap modulation predominates yielding the opposite oscillatory behaviors to the high-field ones, which can be understood by the so-called Doppler shift argument. It is emphasized that the source in the oscillatory behavior depends on the region in the H-T phase diagram. We can safely conclude that the source of the oscillatory behaviors is the modulation of the gap, if a sign change is observed in the oscillatory amplitude of $C(H)$ at low enough temperatures.

6 Conclusion

We have presented a simple calculational scheme of thermodynamic quantities for singlet and unitary triplet states under magnetic fields. A combination of the approximate analytic solution with a free energy functional in the quasiclassical theory provides a wide use formalism including the impurity scattering and the multiband superconductivity. We have discussed the simple formula for the upper critical fields in terms of the microscopic free energy under magnetic fields. The theory requires a little numerical computation as easy as the usual BCS theory without magnetic fields.

We have demonstrated the application to s-wave, $d_{x^2-y^2}$-wave and two-band s-wave cases to compare with reliable numerical calculations. It concludes that the BPT approximation works quantitatively well for almost all region of the H-T phase diagram. The anisotropy of the upper critical fields governs the high-field or high-temperature oscillatory behaviors of C in the oriented magnetic fields. On the other hand, the low-field and the low-temperature behaviors can be understood by the Doppler shift argument, and other sources such as the Fermi velocity anisotropy to alter the conclusion becomes much less important in this region. Since these effects have the opposite contribution, we can safely conclude that the direct contribution from the gap modulation is detected when a sign change of the oscillatory amplitude is observed in $C(H)$. Otherwise, some extrinsic contribution plays a role in high-temperature high-field region. The further applications of the present formalism such as the interplay of the multi-band superconductivity[42, 58, 59] and the field-angle dependence of the magnetization[71, 72] are given in elsewhere.

Acknowledgments

The author would like to thank I. Vekhter, M. Matsumoto, T. Dahm, D.F. Agterberg, M. Sigrist, T.M. Rice, K. Izawa, N. Nakai, Y. Matsuda, I. Mazin, T. Watanabe, K. Deguchi, T. Nomura, M. Mugikura and M. Udagawa for fruitful discussions. A part of the present work was done in the warm hospitality during his stay in the Institut für Theoretische Physik, ETH Zurich, Switzerland. This work was also supported by the NEDO of Japan and Swiss National Fund.

A Circle Product

In this Appendix, we derive the expression of the circle product, which is the Fourier transform, $C(p,s)$, $s = (x_1 + x_2)/2$, with respect to the relative coordinate $x_1 - x_2$ for a given function of the form

$$C(x_1, x_2) = \int dx_3 A(x_1, x_3) B(x_3, x_2). \tag{93}$$

Let us introduce the relative coordinates, $r = x_1 - x_3$ and $r' = x_3 - x_2$. Substituting the Fourier transforms with respect to r and r',

$$A(x_1, x_3) = \int \frac{dp}{2\pi} e^{ipr} A(p, s + r'/2), \tag{94}$$

$$B(x_3, x_2) = \int \frac{dp'}{2\pi} e^{ip'r'} B(p', s - r/2), \tag{95}$$

and expanding with respect to r and r', we have

$$C(x_1, x_2) = \int dx_3 \int \frac{dp}{2\pi} \int \frac{dp'}{2\pi} e^{i(pr+p'r')} \sum_{nm} \frac{1}{n!m!} \times \frac{\partial^n}{\partial R^n} \frac{\partial^m}{\partial R'^m} A(p,R) B(p',R') \left(\frac{r'}{2}\right)^n \left(-\frac{r}{2}\right)^m \Bigg|_{R=R'=} \tag{96}$$

Using the identity, $(ir')^n = \partial^n e^{ip'r'}/\partial p'^n$, we obtain

$$C(x_1, x_2) = \int \frac{dp}{2\pi} e^{ip(x_1-x_2)} \delta_{p,p'} \delta_{R,s} \delta_{R',s} \times \exp\left[\frac{1}{2i}\left(\frac{\partial}{\partial p'}\frac{\partial}{\partial R} - \frac{\partial}{\partial p}\frac{\partial}{\partial R'}\right)\right] A(p,R) B(p',R'). \tag{97}$$

By definition of the Fourier transform, the desired expression is

$$C(p, s) = \delta_{p,p'} \delta_{R,s} \delta_{R',s} \times \exp\left[\frac{1}{2i}\left(\frac{\partial}{\partial p'}\frac{\partial}{\partial R} - \frac{\partial}{\partial p}\frac{\partial}{\partial R'}\right)\right] A(p,R) B(p',R'). \tag{98}$$

B Properties of Faddeeva Function

Here we summarize the properties of the Faddeeva function. The FORTRAN code of the function may be found at the URL[73]. The Faddeeva function or the normalized error function is defined by

$$W(z) = e^{-z^2} \mathrm{erfc}(-iz) = \frac{i}{\pi} \int_{-\infty}^{\infty} dt \frac{e^{-t^2}}{z - t}, \tag{99}$$

for $\mathrm{Im}\, z > 0$. From this expression, we immediately have $W(-z^*) = [W(z)]^*$ and it is a real quantity for a pure imaginary argument, $z = ix$. It is easy to show that the even derivative of $W(z)$ is also real for $z = ix$ and the odd derivative is pure imaginary. The first derivative of $W(z)$ is given by $W^{(1)}(z) = -2zW(z) + 2i/\sqrt{\pi}$. The m-th derivative ($m \geq 2$) satisfies the recursion formula:

$$W^{(m)}(z) = -2zW^{(m-1)}(z) - 2(m-1)W^{(m-2)}(z). \tag{100}$$

We can express $W(z)$ as a continued fraction,

$$W(z) = \frac{i}{\sqrt{\pi}} \frac{1}{z-} \frac{1/2}{z-} \frac{1}{z-} \cdots \tag{101}$$

Then, the asymptotic expression is obtained for $|z| \gg 1$ as

$$W(z) = \frac{i}{\sqrt{\pi} z} \left(1 + \frac{1}{2z^2} + \cdots \right), \tag{102}$$

$$W^{(1)}(z) = -\frac{i}{\sqrt{\pi} z^2} \left(1 + \frac{3}{2z^2} + \cdots \right). \tag{103}$$

The Taylor expansion of $W(z)$ is $\sum_{n=0}^{\infty} (iz)^n / \Gamma(n/2+1)$. The explicit expression for $|z| \ll 1$ is

$$W(z) = 1 - z^2 + \frac{2i}{\sqrt{\pi}} z \left(1 - \frac{2}{3} z^2 + \cdots \right), \tag{104}$$

$$W^{(1)}(z) = -2z(1 - z^2) + \frac{2i}{\sqrt{\pi}} \left(1 - 2z^2 + \cdots \right). \tag{105}$$

References

[1] Leggett, A.J.; *Rev. Mod. Phys.* 1975, 47, 331.

[2] Sigrist, M.; Ueda, K. *Rev. Mod. Phys.* 1991, 63, 239.

[3] Vekhter, I.; Hirschfeld, P.J.; Nicol, E.J. *Phys. Rev. B* 2001, 64, 064513.

[4] Joynt, R.; Taillefer, L. *Rev. Mod. Phys.* 2002, 74, 235.

[5] Mackenzie, A.P.; Maeno, Y. *Rev. Mod. Phys.* 2003, 75, 657.

[6] Choi, H.J.; Roundy, D.; Sun, H.; Cohen, M.L.; Louie, S.G. *Nature* (London) 2002, 418, 758.

[7] Ohashi, Y. *J. Phys. Soc. Jpn.* 2002, 71, 1978.

[8] Koshelev, A.E.; Golubov, A.A. *Phys. Rev. Lett.* 2003, 90, 177002.

[9] Miranović, P.; Machida, K.; Kogan, V.G. *J. Phys. Soc. Jpn.* 2003, 72, 221.

[10] Dahm, T.; Schopohl, N. *Phys. Rev. Lett.* 2003, 91, 017001.

[11] Zhitomirsky, M.E.; Dao, V.-H. *Phys. Rev. B* 2004, 69, 054508.

[12] Bouquet, F.; Wang, Y.; Sheikin, I.; Plackowski, T.; Junod, A.; Lee, S.; Tajima, S. *Phys. Rev. Lett.* 2002, 89, 257001.

[13] Nakai, N.; Ichioka, M.; Machida, K. *J. Phys. Soc. Jpn.* 2002, 71, 23.

[14] Sologubenko, A.V.; Jun, J.; Kazakov, S.M.; Karpinski, J.; Ott, H.R. *Phys. Rev. B* 2002, 66, 014504.

[15] Kusunose, H.; Rice, T.M.; Sigrist, M. *Phys. Rev. B* 2002, 66, 214503.

[16] Izawa, K.; Yamaguchi, H.; Matsuda, Y.; Shishido, H.; Settai, R.; Onuki, Y. *Phys. Rev. Lett.* 2001, 87, 057002.

[17] Izawa, K.; Takahashi, H.; Yamaguchi, H.; Matsuda, Y.; Suzuki, M.; Sasaki, T.; Fukase, T.; Yoshida, Y.; Settai, R.; Onuki, Y. *Phys. Rev. Lett.* 2001, 86, 2653.

[18] Tanatar, M.A.; Suzuki, M.; Nagai, S.; Mao, Z.Q.; Maeno, Y.; Ishiguro, T. *Phys. Rev. Lett.* 2001, 86, 2649.

[19] Tanatar, M.A.; Nagai, S.; Mao, Z.Q.; Maeno, Y.; Ishiguro, T. *Phys. Rev. B* 2001, 63, 064505.

[20] Izawa, K.; Kamata, K.; Nakajima, Y.; Matsuda, Y.; Watanabe, T.; Nohara, M.; Takagi, H.; Thalmeier, P.; Maki, K. *Phys. Rev. Lett.* 2002, 89, 137006.

[21] Izawa, K.; Yamaguchi, H.; Sasaki, T.; Matsuda, Y. *Phys. Rev. Lett.* 2002, 88, 027002.

[22] Deguchi, K.; Mao, Z.Q.; Yaguchi, H.; Maeno, Y. *Phys. Rev. Lett.* 2004, 92, 047002.

[23] Aoki, H.; Sakakibara, T.; Shishido, H.; Settai, R.; Onuki, Y.; Miranovic, P.; Machida, K. *J. Phys. Condens. Matter* 2004, 16, L13.

[24] Park, T.; Chia, E.E.M.; Salamon, M.B.; Bauer, E.D.; Vekhter, I.; Thompson, J.D.; Choi, E.M.; Kim, H.J.; Lee, S.-I.; Canfield, P.C. *Phys. Rev. Lett.* 2004, 92, 237002.

[25] Park, T.; Salamon, M.B.; Choi, E.M.; Kim, H.J.; Lee, S.-I. *Phys. Rev. B* 2004, 69, 054505.

[26] Park, T.; Salamon, M.B.; Choi, E.M.; Kim, H.J.; Lee, S.-I. *Phys. Rev. Lett.* 2003, 90, 177001.

[27] Deguchi, K.; Mao, Z.Q.; Maeno, Y. *J. Phys. Soc. Jpn.* 2004, 73, 1313.

[28] Tinkham, M. *Introduction to superconductivity*, McGraw-Hill, NY, 1975.

[29] Volovik, G.E. *Pis'ma Zh. Éksp. Teor. Fiz.* 1993, 58, 457 [*JETP Lett.* 1993, 58, 469].

[30] Franz, M.; Tesanovic, *Z. Phys. Rev. B* 1999, 60, 3581.

[31] Vekhter, I.; Hirschfeld, P.J.; Carbotte, J.P.; Nicol, E.J. *Phys. Rev. B* 1999, 59, R9023.

[32] Won, H.; Maki, K. *Europhys. Lett.* 2001, 54, 248.

[33] Dahm, T.; Graser, S.; Iniotakis, C.; Schopohl, N. *Phys. Rev. B* 2002, 6, 144515.

[34] Klein, U. *J. Low Temp. Phys.* 1987, 69, 1.

[35] Pöttinger, B.; Klein, U. *Phys. Rev. Lett.* 1993, 70, 2806.

[36] Ichioka, M.; Hayashi, N.; Machida, K. *Phys. Rev. B* 1997, 55, 6565.

[37] Ichioka, M.; Hasegawa, A.; Machida, K. *Phys. Rev. B* 1999, 59, 8902.

[38] Ichioka, M.; Machida, K. *Phys. Rev. B* 2002, 65, 224517.

[39] Yu, F.; Salamon, M.B.; Leggett, A.J.; Lee, W.C.; Ginsberg, D.M. *Phys. Rev. Lett.* 1995, 74, 5136.

[40] Vekhter, I.; Houghton, A. *Phys. Rev. Lett.* 1999, 83, 4626.

[41] Udagawa, M.; Yanase, Y.; Ogata, M. *Phys. Rev. B* 2004, 70, 184515.

[42] Kusunose, H. *J. Phys. Soc. Jpn.* 2004, 73, 2512.

[43] Miranović, P.; Nakai, N.; Ichioka, M.; Machida, K. *Phys. Rev. B* 2003, 68, 052501.

[44] Brandt, U.; Pesch, W.; Tewordt, L. *Z. Phys.* 1967, 201, 209.

[45] Pesch, W. *Z. Phys. B: Condens. Matter* 1975, 21, 263.

[46] Eilenberger, G. *Z. Phys.* 1968, 214, 195.

[47] Larkin, A.I.; Ovchinnikov, Y.N. *Zh. Exp. Teor. Fiz.* 1968, 55, 2262 [*Sov. Phys. JETP* 1969, 28, 1200].

[48] Kusunose, H. *Phys. Rev. B* 2004, 70, 054509.

[49] Klimesch, P.; Pesch, W. *J. Low Temp. Phys.* 1978, 32, 869.

[50] Serene, J.W.; Rainer, D. *Phys. Rep.* 1983, 101, 221.

[51] Rammer, J.; Smith, H. *Rev. Mod. Phys.* 1986, 58, 323.

[52] Houghton, A.; Vekhter, I. *Phys. Rev. B* 1998, 57, 10831.

[53] Eckern, U.; Schmid, A. *J. Low Temp. Phys.* 1981, 45, 137.

[54] Shiba, H. *Prog. Theor. Phys.* 1968, 40, 435.

[55] Schmitt-Rink, S.; Miyake, K.; Varma, C.M. *Phys. Rev. Lett.* 1986, 57, 2575.

[56] Hirschfeld, P.J.; Vollhardt, D.; Wölfle, P. *Solid State Commun.* 1986, 59, 111.

[57] Rieck, C.T.; Scharnberg, K.; Schopohl, N.J. *J. Low Temp. Phys.* 1991, 84, 381.

[58] Kaur, R.P.; Agterberg, D.F.; Kusunose, H. *cond-mat/0503289*.

[59] Mugikura, M.; Kusunose, H.; Kuramoto, Y. *cond-mat/0412522*.

[60] Suhl, H.; Matthias, B.T.; Walker, L.R. *Phys. Rev. Lett.* 1959, 3, 552.

[61] Kondo, J. *Prog. Theor. Phys.* 1963, 29, 1.

[62] Leggett, A.J. *Prog. Theor. Phys.* 1966, 36, 901.

[63] Kulić, M.L.; Dolgov, O.V. *Phys. Rev. B* 1999, 60, 13062.

[64] Ichioka, M.; Machida, K.; Nakai, N.; Miranović, P. *Phys. Rev. B* 2004, 70, 144508.

[65] Eskildsen, M.R.; Jenkins, N.; Levy, G.; Kulger, M.; Fischer, Ø.; Jun, J.; Kazakov, S.M.; Karpinski, J. *Phys. Rev. B* 2003, 68, 100508.

[66] Miranović, P.; Ichioka, M.; Machida, K. *Phys. Rev. B* 2004, 70, 104510.

[67] Nakai, N.; Miranović, P.; Ichioka, M.; Machida, K. *Phys. Rev. B* 2004, 70, 100503.

[68] Although the numerical results are obtained by using gap with six line nodes, the field dependence of the zero-energy DOS almost coincides with that of $d_{x^2-y^2}$-wave (N. Nakai, private communication).

[69] Brandt, E.H. *Phys. Rev. B* 2003, 68, 054506.

[70] Miyake, K.; Narikiyo, O. *Phys. Rev. Lett.* 1999, 83, 1423.

[71] Adachi, H.; Miranović, P.; Ichiokda, M.; Machida, K. *Phys. Rev. Lett.* 2005, 94, 067007.

[72] Kusunose, H. *J. Phys. Soc. Jpn.* 2005, 74, 1119.

[73] *http://www.acm.org/toms/V16.html*

In: Superconductivity, Magnetism and Magnets
Editor: Lannie K. Tran pp. 31-53
ISBN 1-59454-845-5

Chapter 2

THE CRITICAL FIELD OF A TYPE II SUPERCONDUCTING FILAMENTARY STRUCTURE IN A SPATIALLY INHOMOGENEOUS APPLIED MAGNETIC FIELD

M. Masale
Physics Department, University of Botswana,
P/Bag 0022, Gaborone, Botswana

Abstract

A review is given of the calculation of surface nucleation of superconductivity in low-temperature type II superconductors with cylindrical symmetry based on the linearized Ginzburg-Landau (G-L) equation for the order parameter. Two distinct configurations of a spatially inhomogeneous applied magnetic field considered are when the magnetic field is applied either parallel to the axis of the superconducting structure or in the azimuthal direction. In the parallel configuration, the universal temperature-field ($\varepsilon - f$) curves are characterized by flux-entry points at each of which the azimuthal quantum number m decreases by unity. In this case, critical field corresponds strictly to zero axial wave number k_z. The quasi-periodicity of the flux-entries increases in f with the increasing degree of the spatial inhomogeneity of the axial applied magnetic field. An increase of the degree of the spatial inhomogeneity of the applied magnetic fiel d is also found to drastically reduce the amplitude of Little-Parks oscillations characteristic of a thin-walled hollow cylinder. The situation is the other way around in the azimuthal configuration, that is $m = 0$ and $k_z \neq 0$, although in general k_z is a continuous variable. Spatial distributions of the superconducting phases are investigated by studying variations of the order parameter as well as the current density across thick specimen. The magnetic field profile, in particular, in the parallel configuration, can be tailored in such a way as to counteract spatial variations of the superconducting phase particularly across a thick specimen. This has the advantage that in more complex evaluations beyond merely a calculation of the critical field, the specimen may then be described in terms of a *constant* wave function.

1 Introduction

It has long been known that superconductivity can be destroyed by an applied magnetic field [1], the flux of the applied magnetic field consequently expelled from a bulk superconductor for fields less than the bulk critical value H_{c2} [2]. At this value of the applied field, a type II superconductor undergoes a second order transition from the superconducting to the normal state. A higher surface nucleation field, $H_{c3} = 1.69H_{c2}$, appears at the planar surface of a type II superconductor in a parallel magnetic field [3]. In their extensive investigations, both experimental and theoretical, Guyon *et al.* [4] found that the surface nucleation field of an ultra-thin film is very different from that of either a thick film or a semi-infinite specimen. At a certain critical thickness of the film, at least in principle, it becomes possible for a vortex to fit inside the film. This phenomenon, characterized by a point of inflection on the temperature-field curve, is associate d with the first flux-entry point. In superconducting structures of the nanoscale subjected to a uniform parallel applied magnetic field, the nature of the superconducting state just below H_{c3} depends on the dimensionless ratio of the specimen thickness to $\xi(T)$, the G-L coherence length. Motivated by how the parallel surface nucleation field changes mainly according to the shape of a superconductor as well as by the advancement in fabrication techniques [5], Masale *et al.*[6] reconsidered a number of systems with cylindrical symmetry. Distinct flux-entry points, which are quantized in integral multiples of the elementary flux quantum, $\phi_o(= h/2e)$, are clearly seen in the universal field-temperature curve of a solid cylinder [7,8] and a hole [6] in a parallel magnetic field. As such, in a very thin-walled hollow cylinder immersed in a parallel magnetic field, the flux-entry points are such a prominent feature of the field-temperature curve giving rise to what are known as Little-Parks oscillations [9]. In systems of the nanoscale, nature of the superconducting state just below H_{c3} depends in a crucial way on the ratio of the system-thickness to the cyclotron radius of a charged particle in a uniform parallel applied magnetic field. The uniform-field cyclotron radius is therefore very useful as a scaling length for the description of surface nucleation.

In reality, however, even in the most refined experimental settings, the applied magnetic field will exhibit some degree of spatial inhomogeneity. There is now a growing interest in the basic properties of superconducting structures subjected to a static but spatially inhomogeneous applied magnetic field [10,11]. Clearly, a non-uniform applied magnetic field can no longer be uniquely characterized in terms of the cyclotron radius. This then raises the question of what scaling parameter to use instead of the uniform-field cyclotron radius. In fact, the orbits of a charged particle under the influence of a non-uniform magnetic field are predominantly snake-like in character [12,13] and are therefore not closed for a wide range of the relevant parameters, if at all for any [14]. Although following a rather simplistic approach based on the linearized Ginzburg-Landau equation, the results obtained by Masale [15] suggest that there may be possible advantages to be gained from the applic ation of a spatially inhomogeneous magnetic field. The profile of the applied magnetic field may be tailored in such a way as to counteract spatial variations of the superconducting phase particularly across a thick specimen. Nucleation of superconductivity should then be uniform even in a thick specimen and a complete description of the specimen can then be given in terms of a *constant* ground state wave function. In principle, this should help ease mathematical complexities particularly in calculations where otherwise it is necessary to employ

the full G-L equation for the order parameter [11,16,17]. While conventional experiments provide information on the onset of superconductivity [18,19], the above mentioned spatial variations of the superconducting phases should easily be accessible in the newly developed micro-magnetization measurement technique [20].
In systems with plano surfaces, nucleation of superconductivity occurs at a much lower temperature for the perpendicular than for the same value of the parallel applied magnetic field. It is therefore naturally of interest to investigate how the fundamental properties of superconducting filaments are influenced by the orientation of the applied magnetic field relative to the dimensions of system of different geometries. There have recently been some investigations of a type II superconducting cylindrical filament subjected to an azimuthal applied magnetic field [21, 22]. The rapid advance in nanofabrication technologies to produce very fine wires based on high-T_c materials opens avenues for potential applications in circuitry devices. As part of the circuitry, it would be crucial to determine the regimes of relevant parameters, in particular, the current density, for which such wires are either in the normal or the superconducting phase. Now, a current of uniform density, as in the investigations mentioned above, produces an inherently spatially inhomogeneous magnetic field. A further complication in this configuration of the non-uniform applied magnetic field is that the snake-like orbits of the charges are in the plane containing the axis of the cylindrical specimen. The calculation of the magnetic flux enclosed within the charged particle's orbit is inherently not easily tractable.
The main objective of this paper is to give a brief overview of the nature of surface nucleation of superconductivity in low-temperature type II superconductors with cylindrical symmetry. The general problem of a superconducting cylindrical filament subjected to a magnetic field with two components; the axial and the azimuthal; is separable. However since any co-planar vector can be resolved into two perpendicular components, it is convenient to evaluate the influence of each component separately. In particular, the two distinct configurations considered are when the applied magnetic field is either parallel to the axis of the superconducting structure or in the azimuthal direction. Cladding of a superconductor with a suitable normal metal is often used in the suppression of unwanted surface nucleation. A detailed account of this effect may be found in some earlier investigations [6,11,21 23] and is not discussed here for brevity.

2 Theory

For the calculation of the nucleation field of a type II superconductor, it is sufficient to employ the linearized G-L equation for the order parameter, ψ, given by [24]:

$$\frac{1}{2\mu}(-\iota\hbar\nabla - 2e\mathbf{A})^2\psi + \alpha\psi = 0, \tag{1}$$

where $\mathbf{A}$ is the vector potential, μ is the mass of a particle of charge $2e$ and $\alpha = \alpha_o(T - T_{cb})$, essentially the temperature T, is the G-L parameter in terms of the bulk critical value T_{cb}. In view of the symmetry of the problems posed here, the solution of the linearized G-L equation is sought in the general form:

$$\psi = C_{m\ell}\exp(ik_z z)\exp(\iota\varphi)\chi(\rho); \quad m = 0, \pm1, \pm2...., \tag{2}$$

where $C_{m\ell}$ is a constant, k_z is the axial wavenumber, and ℓ and m are the radial and azimuthal quantum numbers, respectively. The superconducting phase of a specimen may be inferred from spatial variations of the order parameter or the so-called supercurrent [21]. The general expression for the quantum mechanical current density, or supercurrent, is given by

$$\mathbf{J} = -(\iota e\hbar/\mu)(\psi^*\nabla\psi - \psi\nabla\psi^*) - (4e^2/\mu)|\psi|^2\mathbf{A}. \tag{3}$$

In the case of a free-standing superconductor or a superconductor-insulator interface, the geometry of the superconductor enters the eigenvalue problem via the boundary condition [23]:

$$\hat{\mathbf{n}} \cdot (-\iota\hbar\nabla - 2e\mathbf{A})\psi = 0, \tag{4}$$

where $\hat{\mathbf{n}}$ is a unit vector normal to the interface.
The theory outlined in this section is now applied to specific systems which poses cylindrical symmetry in the subsequent sections. As stated earlier, the externally applied magnetic field is either in the axial or azimuthal direction.

3 Parallel Applied Magnetic Field

The most general system in this configuration of the applied magnetic field is a type II superconducting hollow cylinder of thickness $(R_2 - R_1)$ and length L_z. The spatially inhomogeneous static applied magnetic field, which satisfies all the magnetostatic equations, taken with only the radial and the axial components, is given by

$$B_\rho = \frac{B_\lambda}{\rho}(|z| - \frac{1}{2}L_z) \quad \text{and} \quad B_z = B_\mu + B_\lambda \ln(\rho/R_1), \tag{5}$$

respectively, where B_μ and B_λ are constants measured in the obvious appropriate units. The vector potential $\mathbf{A} = (0, A_\varphi, A_z)$ of the applied magnetic field is taken in the gauge:

$$A_\varphi = \frac{1}{2}\bar{B}\rho[\nu - \frac{1}{2}\sigma + \frac{1}{2}\sigma\ln(\rho^2/R_1^2)] \quad \text{and} \quad A_z = \bar{B}\sigma(|z| - \frac{1}{2}L_z)\varphi, \tag{6}$$

in which

$$\bar{B} = (B_\mu + B_\lambda), \;\; \nu = B_\mu/\,\bar{B} \text{ and } \sigma = B_\lambda/\,\bar{B}. \tag{7}$$

It is worth commenting that with the singularity of the above magnetic field at the origin smoothed out, for example by introducing a tiny core at the origin, this profile of the applied magnetic field mimics that just near the ends of a magnetic dipole. This is particularly the case for a core of the electromagnet fashioned to have a conical end-face. A thought experimental set-up is then one of a mesoscopic structure placed near the end-face of a magnetic dipole such that the axes of the mesoscope and that of the core of the magnet exactly coincide. In this experimental set-up, it is anticipated that surface nucleation effects due B_ρ will be negligibly small, more so that $B_\rho = 0$ at $z = \pm L_z/2$, anyway. Further, an increase of B_λ should give rise to an expanding normal phase of the inner region at $z = \pm L_z/2$; the expansion centered at $\rho = 0$. Now, even with the influence of B_ρ neglected, the axial component of the field constitutes two surface nucleation fields: $B_{c\perp}$; the nucleation field perpendicular to the end-faces and B_{c3}; the critical field parallel to the cylinder-axis.

As stated earlier, it is well established that nucleation of superconductivity occurs at a much lower temperature for the perpendicular than for the same value of the parallel appl ied magnetic field. For this reason, $B_{c\perp}$ can be neglected over its parallel counterpart, B_{c3}. In view of these considerations, the problem considered here essentially reduces to that of a superconducting mesoscopic structure subjected only to the spatially inhomogeneous axial magnetic field $\mathbf{B} = B_z\hat{\mathbf{z}}$. Now, in order to avoid complicated profiles of the applied field, for example, a change of sign of the applied field across the thickness of the superconducting shell, the constraint: $\nu + \sigma = 1$: is imposed on the values of ν and σ, both assumed to be positive. With this constraint, the results obtained here then directly relate and reduce to the special case of a cylinder in a uniform axial applied magnetic field [6-9,25]. As stated earlier, a spatially inhomogeneous magnetic field can not be represented uniquely by the cyclotron parameters of the radius and frequency. It is nonetheless convenient, here, to define fictitious "cyclotr on" quantities of the radius and frequency as $\bar{a}_c = (\hbar/e\bar{B})^{1/2}$ and $\bar{\omega}_c = e\bar{B}/\mu$, respectively, that is, in terms of the value of the magnetic field at a radial distance ρ_e such that $\ln(\rho_e/R) = 1$. Analogous to the problem of a hollow cylinder in a uniform parallel applied magnetic field [6], the temperature ε_i and the field $\bar{f}_i$ are defined by the following dimensionless variables

$$\varepsilon_i = \mu|\alpha|R_i^2/2\hbar^2 \text{ and } \bar{f}_i = R_i^2/4\bar{a}_c^2 \text{ with } i = 1 \text{ or } 2. \tag{8}$$

With the specific form the component A_φ of the vector potential given by equation (6), the G-L equation for the order parameter in the radial function $\chi(\rho)$ becomes

$$\rho\frac{d}{d\rho}\left(\rho\frac{d\chi}{d\rho}\right) = \left\{[2\frac{\mu}{\hbar^2}\alpha + k_z^2]\rho^2 + \left[m + \frac{e}{\hbar}\bar{B}\rho^2(\nu - \frac{1}{2}\sigma + \frac{1}{2}\sigma\ln(\rho^2/R_1^2))\right]^2\right\}\chi. \tag{9}$$

It is not possible, in particular, because of the logarithmic terms, to cast equation (9) into a canonical form. However, solutions of this equation may be found in closed form if the logarithmic terms are replaced by their linear fitting forms as follows:

$$\ln\varsigma \approx c_1 + c_2\varsigma \quad \text{and} \quad \ln^2\varsigma \approx d_1 + d_2\varsigma. \tag{10}$$

One simple procedure for obtaining the constant factors c_i and d_i ($i = 1$ or 2) is as follows. First, plot graphs of the exact logarithmic terms as functions of the argument ς using standard graphical software, for example, such as MSDOS-grapher. Next, perform a polynomial fitting, in general, of degree N on each curve of the logarithmic terms. On performing the fitting, the software package automatically computes and lists the relevant various constant factors appropriate to the particular range of ς and corresponding to any degree N of the fitting. It is worth commenting that a much closer fit is found for the second ($\ln^2\varsigma$) of these terms. Now, employing only the first of the approximate linear representation of the logarithmic functions given by equation (10), the substitution

$$\chi = \zeta^{|m|/2}e^{-\zeta/2}\mathcal{F}, \tag{11}$$

where

$$\zeta = \left[\nu - \frac{1}{2}(c_1 + 1 + \ln\eta)\sigma\right] \times \rho^2/2\bar{a}_c^2, \tag{12}$$

in which $\eta = R_1^2/R_2^2$, leads to

$$\zeta \frac{d^2\mathcal{F}}{d\zeta^2} + (b - \zeta)\frac{d\mathcal{F}}{d\zeta} - a\mathcal{F} = 0, \tag{13}$$

which is the canonical form of Kummer's equation for the confluent hypergeometric function. The complete solution of equation (13) may therefore be written as a linear combination of the M and U confluent hypergeometric functions [6]:

$$\psi = \zeta^{|m|/2} \exp(-\zeta/2)\Big\{PM(a,b,\zeta) + QU(a,b,\zeta)\Big\} \exp(\iota m\varphi) \exp(\iota k_z z), \tag{14}$$

where P and Q are constants and the parameters a and b are given by

$$a = \frac{1}{2} + \frac{1}{2}|\lambda_1| + \frac{1}{2}\lambda_1 - \frac{1}{2}\frac{\varepsilon_2}{\lambda_2 \bar{f}_2} + \frac{1}{4}\frac{k_z^2 R_2^2}{\lambda_2 \bar{f}_2} \text{ and } b + |\lambda_1| + 1, \tag{15}$$

in which

$$\lambda_1 = m - c_2\sigma \bar{f}_2 \text{ and } \lambda_2 = \nu - \frac{1}{2}(c_1 + 1 + \ln\eta)\sigma. \tag{16}$$

The application of the insulator-superconductor boundary condition at the surfaces $\rho = R_1$ and $\rho = R_2$ of the free-standing mesoscope, which in this case simplifies to $\frac{d\psi}{d\rho} = 0$, leads to the following eigenvalue equation:

$$\mathcal{D}_M(2\bar{f}_1)\mathcal{D}_U(2\bar{f}_2) - \mathcal{D}_M(2\bar{f}_2)\mathcal{D}_U(2\bar{f}_1) = 0, \tag{17a}$$

in which

$$\mathcal{D}_M(2\bar{f}_i) = \frac{1}{2}\Big[|\lambda_1| - 2\bar{f}_i\Big] M(a,b,2\bar{f}_i) + \frac{2a\bar{f}_i}{b} M(a+1,b+1,2\bar{f}_i) \tag{17b}$$

and

$$\mathcal{D}_U(2\bar{f}_i) = \frac{1}{2}\Big[|\lambda_1| - 2\bar{f}_i\Big] U(a,b,2\bar{f}_i) - 2a\bar{f}_i U(a+1,b+1,2\bar{f}_i). \tag{17c}$$

For a given set of the relevant parameters, generating the field-temperature curves is essentially a matter of standard root-finding routine. First, fix $\bar{f}$ then solve the eigenvalue equation, searching for the negative fluxoid number m which gives the lowest value of ε. Only the minimum value of ε, of course corresponding to $k_z = 0$, has physical significance since it is one that corresponds to the required critical field. However, as demonstrated in the investigations by Masale [11], perhaps a more appropriate representation of the spatially inhomogeneous magnetic field is in terms of the actual (approximate) flux enclosed within the outer surface R_2 of the hollow cylinder given by

$$f = \frac{\Phi}{2\phi_o} = \frac{1}{2\phi_o}\int_0^R B \cdot dS = \Big[\nu + (c_1' + \frac{2}{3}c_2' - \ln\eta)\sigma\Big]\bar{f}_2, \tag{18}$$

where the constants c_i' correspond to the range $0 \leq \rho \leq R_2$ of the radial distance. Representing the applied magnetic field even as f, in particular, for the calculation of H_{c3}, somewhat overshadows its (magnetic field) spatial inhomogeneity. This is because the specimen should respond exactly the same way to any form of the magnetic field that produces the magnetic flux, Φ, prescribed by equation (18).

Now, in the case of a free-standing superconducting mesoscope, the order parameter assumes a constant value, at least at the surface of the superconductor, in keeping with the boundary conditions prescribed by equation (4). Further, there can hardly be any phase changes of the order parameter across a specimen of small thickness, less than or equal to the G-L coherence length. This is in spite of smooth spatial inhomogeneity of the applied magnetic field [11,21]. As in earlier analyses [6,25], the limiting form of ε_2 for small fields may be obtained by integrating the linearized G-L equation for the order parameter across the shell-thickness. Nucleation of superconductivity may then be adequately described by the following so-called constant ψ approximate result:

$$\varepsilon_2 = \frac{1}{2}m^2\frac{1}{1+\eta} - g_1|m|\bar{f}_2 + g_2\bar{f}_2^2, \tag{19a}$$

where

$$g_1 = \left[\nu - \frac{3}{4}\sigma - \frac{1}{2}\sigma\frac{\ln\eta}{[1-\eta^2]}\right] \tag{19b}$$

and

$$g_2 = \frac{2}{3}\frac{[1-\eta^3]}{[1-\eta^2]}\left[[\nu - 2\sigma/3]^2 + [\sigma/6]^2 - \frac{\sigma\ln\eta}{[1-\eta^3]}[\nu - \frac{2}{3}\sigma - \sigma\ln\eta]\right]. \tag{19c}$$

The constant ψ result can also be useful as a check on the 'full' numerical result in the case of less tractable problems [11,21].

The general expression for the quantum mechanical current density in this case takes the more explicit approximate form:

$$J_\varphi/J_{o\varphi} = -\left\{|m|/x + 2\bar{f}_2 x\left[\nu - \frac{1}{2}\sigma(1+\ln\eta - c_1 - c_2x^2)\right]\right\}\chi^2, \tag{20}$$

where $x = \rho/R_2$ and $J_{o\varphi} = 2e\hbar/(\mu R_2)$.

3.1 Free-Standing Disk

This is the system of a solid cylinder of radius R and length L_z subjected to a parallel applied magnetic field $\mathbf{B} = \mathbf{B}(\rho)\hat{\mathbf{z}}$, given by equation (5). Application of the insulator-superconductor boundary condition leads to the following eigenvalue equation

$$\mathcal{D}_M(2f) = \frac{1}{2}\left[|\lambda_1| - 2f\right]M(a,b,2f) + \frac{2af}{b}M(a+1,b+1,2f_2). \tag{21}$$

The subscripts on the critical temperature and field variables have been dropped for convenience. The above result is in fact a special case of the general eigenvalue equation (17), in the limit of a very tiny core $R_1 \to 0$ and with the logarithmic term, $\ln\eta$, identically set equal to zero.

Figure 1 shows the universal temperature-field curves of a solid cylinder; as plots of ε versus f; some few values of indices of the degree of the magnetic field inhomogeneity; $\sigma(= 1-\nu)$. Each curve corresponds to a different value of ν, at say $f = 5.0$, as follows:

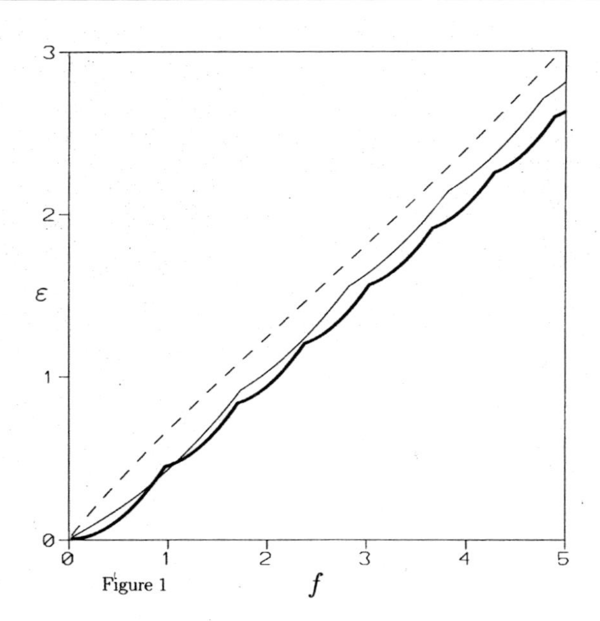

Figure 1: The universal field-temperature ($\varepsilon - f$) curves of a type II superconducting disk in a spatially inhomogeneous axial magnetic field. Each curve corresponds to a different value of ν as follows: $\nu = 0.6$ for the dashed curve, $\nu = 0.8$ for the thin solid line and $\nu = 1.0$, the special case of a uniform magnetic field, for thick solid line.

$\nu = 1.0$ for the lowest (thick) curve, $\nu = 0.8$ for the (middle) smooth curve and $\nu = 0.6$ for the dashed (highest) curve. Note that the thick curve is the result for the special case of a uniform applied magnetic field; $\nu = 1.0$. The constants, corresponding to the range $\sim 0 \leq \rho \leq R$ of the linear representation of the logarithmic factor given by equation (10) are $c_1' = -2.59$ and $c_2' = 3.14$. The general trend particularly for near-uniform applied magnetic field, $\nu \sim 1$, is that each curve consists of joint segments each corresponding to unit stepwise decrease of m as the field increases [11]. It is seen, however, a decrease of ν leads to a dramatic increas e of the periodicity in f of the cusps. In fact the cusps are completely wiped out for values of $\nu \leq 0.5$. This effect as well as the apparent suppression of the critical temperature are less emphasized in the results of an earlier investigation for the same system but where a slightly different approximation for the applied magnetic field was employed [11].

3.2 Free-Standing Short Hole

The system considered here is a cylindrical cavity of radius R and length L_z "dug-out" of a sheet of a type II superconducting material, which is a special case of a mesoscope in the limit when $R_2 \rightarrow \infty$. This is perhaps the system of choice for mapping out any distributions of the superconducting phases since the thickness of the specimen is much larger that the G-L coherence length. Again, with the logarithmic term appearing in the previous relevant equations is formally taken to be zero, the eigenvalue equation is explicitly given by

$$\mathcal{D}_U(2\bar{f}) = \frac{1}{2}\Big[|\lambda_1| - 2\bar{f}\Big]U(a,b,2\bar{f}) - 2a\bar{f}U(a+1,b+1,2\bar{f}). \tag{22}$$

Here, as in previous investigations [6], the field is represented in terms of the so-called missing flux, that is, the flux penetrating the axial cross-sectional area of the cavity.

Figure 2 illustrates the universal temperature-field curves of a hole corresponding to exactly the same values of ν as for figure 1. These are: $\nu = 1.0$ for the lowest (thick) curve, $\nu = 0.8$ for the middle smooth curve and $\nu = 0.6$ for the dashed curve. Numerical values of the constant factors used are: $c = -0.112$, $c_2 = 0.373$, $c_1' = -2.59$ and $c_2' = 3.14$, corresponding to the ranges $1 \leq \rho/R \leq 5$ and $0 \leq \rho/R \leq 1$ of the radial distance, respectively. The first of these ranges corresponds to the region of the smallest radius beyond which the radial function is zero, although for the special case of a uniform applied magnetic field [6]. Again, just like in the case of a hole in a uniform magnetic field [6], each curve consists of cusps each corresponding to unit stepwise decrease of m as the field increases. Here, unlike in the case of a solid cylinder above, the periodicity of the flux-entries is somewhat insensitive to variations of the degree o f the spatial inhomogeneity of the applied magnetic field. In fact the curves corresponding to the different values of ν are more or less merely displaced along the uniform-field universal field-temperature curve. From a comparison of figures 1 and 2, it is seen that surface nucleation is more favourable at a surface with positive (at R_2) than one with negative (at R_1) curvature.
The wave functions shown here almost exactly overlap with those for the corresponding special case of a uniform applied magnetic field and are incidentally insensitive to variations of ν. These are not shown here for brevity but instead only the spatial variation of the current densities, which in any case is very similar to those of the wave functions.

Figure 3 shows the functional form of the current density, that is, a plot of $-(J_\varphi/J_{o\varphi})$ versus ρ/R, for a selection of points on the curve corresponding to $\nu = 0.7$, although not shown here. The other relevant parameters are exactly as for figure 2. The values of f used correspond to the increasing intercepts of the current densities as: 0.05, 0.5 then increasing in steps of 0.5 up to 2.0 for the highest curve (at $\rho/R = 1.0$.) The spatial variation of the current densities illustrated in figure 3 implies that for small fields, nucleation of superconductivity is fairly uniform across the superconductor. This is in spite of the spatial inhomogeneity of the applied magnetic field. The building up of the current densities near the surface for large f is indicative of the persistence of a nucleation sheath only at the surface while the bulk of the superconductor is in the normal state.

3.3 Free-Standing Mesoscope

This is the most general system of the geometries for the configuration of the axial applied magnetic field hence the full formalism outlined in section 3.0 applies. The distinguishing feature of this system, in comparison with the previous two, is that it possesses both positive and negative curvatures, for the interplay surface nucleation.

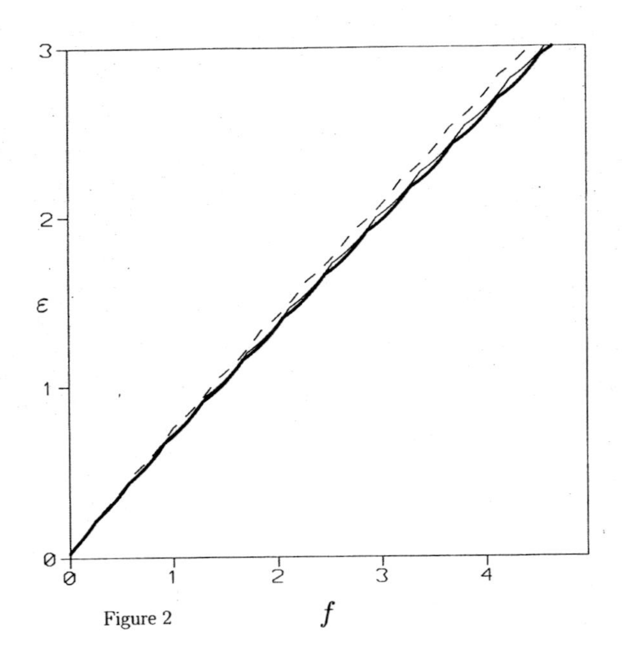

Figure 2: The universal critical field-temperature ($\varepsilon - f$) curves of bulk type II superconductor hosting an empty hole for exactly same values of (ν) as for figure 1. That is, $\nu = 0.6$ for the dashed curve, $\nu = 0.8$ for the thin solid line and $\nu = 1.0$ for thick solid line.

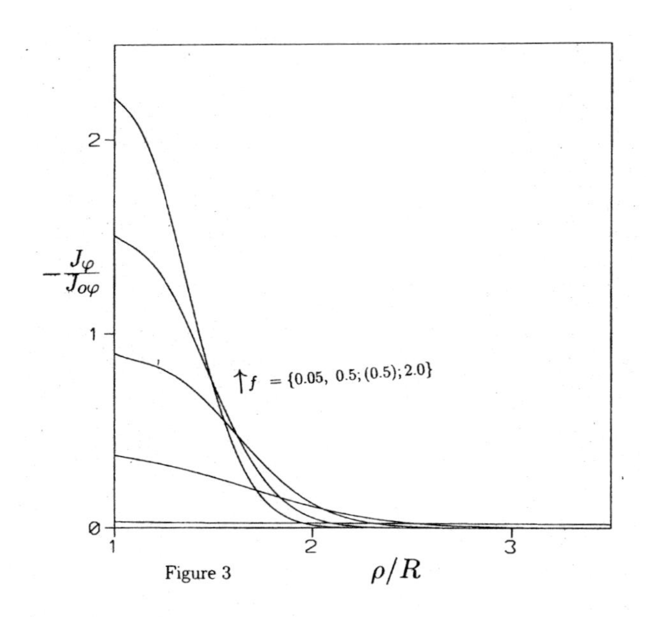

Figure 3: The distribution of the quantum mechanical current density across the superconducting specimen corresponding to $\nu = 0.7$. The values of f used, corresponding to increasing intercepts of $-J_\varphi/J_{o\varphi}$, are: 0.05 (the lowest curve) 0.5 (the next lowest curve) then increasing in steps of $\Delta f = 0.5$ up to $f = 1.0$ (highest curve).

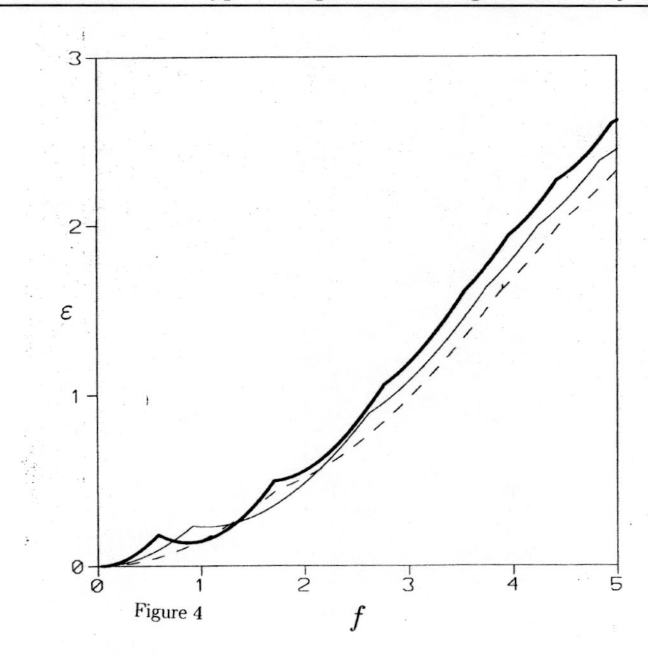

Figure 4: The universal critical field-temperature $\varepsilon_2(=\varepsilon)$ versus $f_2(=f)$ curves of a type II superconducting hollow cylinder of thickness such that $\ln\eta = 1$ in a spatially inhomogeneous axial magnetic field. Each curve corresponds to a different value of ν as follows: $\nu = 0.60$ for the dashed curve, $\nu = 0.80$ for the thin solid line and $\nu = 1.00$ for the thick line.

Figure 4 shows the temperature-field curves of a hollow cylinder of thickness such that $\ln\eta = 1$; as plots of $\varepsilon_2 = \varepsilon$ versus f; for some few values of ν viz: $\nu = 0.60$ the dashed curve, $\nu = 0.8$ the smooth solid curve and $\nu = 1.00$ the thick solid curve. Numerical values of the constant factors used are: $c_1 = -1.63$, $c_2 = 1.74$, $c_1' = -2.59$ and $c_2' = 3.14$, corresponding to the ranges $\sim 0.14 \leq \eta \leq 1$ and $0 \leq \eta \leq 1$ of the radial distance, respectively. As in the case of a uniform applied magnetic field [6,25], the interplay between surface nucleation at the interfaces with positive and negative curvatures of the hollow cylinder gives rise to Little-Parks oscillations. These oscillations are, however, not so pronounced here because of the relatively large thickness of the specimen. As in the case of a superconducting disk, a decrease ν leads to a pronounced increase of periodicity in f of Little-Parks oscillations as well as a reduction of the magnitude of the oscillations. Note again that apart from the apparent wiping out of Little-Parks oscillations, the critical temperature of the hollow cylinder is relatively insensitive to variations of ν, at least in the range of values used. As mentioned earlier, the constant ψ approximate result given by equation (19) can also provide a basis for a more detailed quantitative account of the universal temperature-field curves particularly for a thin-walled hollow cylinder. Such finer details may include, for example, the positions of minima of ε and the periodicity in f of Little-Parks oscillations.

Variations of the radial wave functions across the specimen of thickness such that $\ln\eta \geq 1$ are reminiscent of that of a thick solid cylinder in a uniform applied magnetic field [8] are

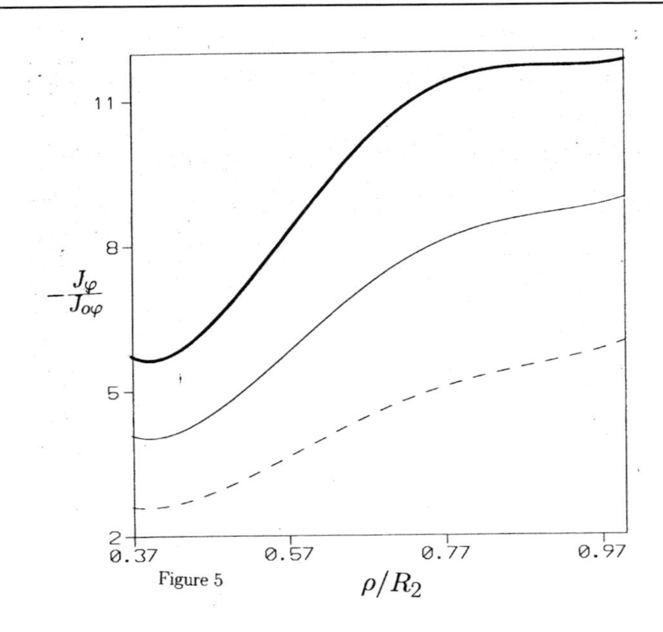

Figure 5: The spatial variations of current densities across a superconducting hollow cylinder of thickness such that $\ln\eta = 1$. The other relevant parameters are $\nu = 0.60$ for the dashed curve, $\nu = 0.7$ for the thin solid line and $\nu = 1.0$ for the thick line.

therefore not shown here for brevity. Such a variation of the wave functions confirms the statement made earlier; that nucleation of superconductivity for this system is favourable at a surface with positive curvature.

Figure 5 shows the functional form of the current density $-(J_\varphi/J_{o\varphi})$ versus ρ/R_2 in a hollow cylinder of thickness such that $\ln\eta = 1$ and $f = 4.0$ for exactly the same values of ν as for figure 4. The values of ν used are: $\nu = 0.50$ for the dashed curve, $\nu = 0.7$ for the smooth thin solid line and $\nu = 1.00$ for the thick curve. The spatial variations of the current densities shown here suggest an increased tightening of nucleation of superconductivity at the outer rather than the inner surface of the hollow cylinder, irrespective of the value of ν.

4 Azimuthal Applied Magnetic Field

This section deals with exactly the same geometries discussed in the previous section except that, here, the applied magnetic field is in the azimuthal direction. In these, the magnetic field is that induced by a current I of uniform density passed along the axis of a cylindrical core. Note that the magnetic field lines are concentric to the radius of the core and are therefore parallel to the surface(s) of the superconducting specimen. Upon increasing the current, the magnetic field reaches a surface-value: $H_{c3}(\phi)$, the azimuthal parallel nucleation field, at which the cylindrical specimen is brought into the normal state. Carrying out indirect resistivity measurements, for example, as in recent experiments on high-T_C superconductors [26], the known critical current can be used in the computation of the required critical field. Although the specific form of the spatially inhomogeneous applied magnetic field depends on the topology of the system in question, surface nucleation of superconductivity is nonetheless described in terms of equations (1) to (3) for a superconductor in a magnetic field.

4.1 Current-Carrying Solid Cylinder

The vector potential of the magnetic field due to a current I of uniform density passed along axis of an infinitely long solid cylinder of radius R is taken in the gauge:

$$A_z = -\frac{1}{2} B_s \frac{\rho^2}{R}, \tag{23}$$

where $B_s = \mu_o I/(2\pi R)$ is the value of the magnetic field at the surface $\rho = R$ of the cylinder and μ_o is the permeability of the specimen, assumed to be non-magnetic. Substitutions of equations (2) and (23) into the G-L equation for the order parameter ψ, and noting that $m = 0$, leads to the following second order differential equation:

$$\varrho \frac{d^2\chi(\varrho)}{d\varrho^2} + \frac{d\chi(\varrho)}{d\varrho} + [\epsilon - (\frac{1}{2}k_z R - f_s \varrho)^2]\chi(\varrho) = 0, \tag{24}$$

where $\varrho = \rho^2/R^2$ and the field $f_s = \pi B_s R^2 / 2\phi_o$ is represented in terms of the surface-value of the magnetic field. As demonstrated in the evaluations by Masale [14], the actual magnetic flux enclosed within the supposedly closed orbits of a charged particle under the influence of this form of the magnetic field is linearly related f_s. It is again a matter of convenience to represent the applied magnetic field as $f = f_s$, that is omitting the subscript on the temperature variable.
Equation (24) above has to be solved numerically since it is not possible to cast it into a canonical form. This makes it essential, therefore, to develop some limiting forms of its solutions which will serve as a guide to the full numerical solution. For a very small current, the term parabolic in ϱ in equation (24) can be neglected and the limiting form of the differential equation that arises is solvable in terms of the confluent hypergeometric function, $M(a, b, \iota\sqrt{|k_z R f_s \varrho})$. An even less numerically taxing approach for generating the field-temperature curve follows from the constant ψ limiting form of the critical temperature for small fields. Integrating equation (24) across the thickness of the cylinder, of course, with the derivatives appearing there set equal to zero, yields the following approximate result:

$$\epsilon \sim \frac{1}{4} k_z^2 R^2 - \frac{1}{2} |k_z| R f + \frac{1}{3} f^2. \tag{25}$$

The series method was adopted for the solutions of equation (24), specifically obtaining the function $\chi(\varrho)$ as well as its first derivative $\chi(\varrho)$, each in terms of a four-term recursion relationship for the coefficients of ϱ. Algorithms were developed for the evaluations of the functions $\chi(\varrho)$ and M based on their series representations. The confidence level regarding the trustworthy of the numerical evaluations of $\chi(\varrho)$ was pivoted on the limiting form that: $\chi(\varrho) \sim M$ as $\varrho \to 0$. Conversion of the series is fast and almost guaranteed for $a < 0$ particularly for small ϱ. It has to be said though that a higher number of terms is required to achieve conversion in the case of $\chi(\varrho)$. Equipped with subroutines for the evaluations of $\chi(\varrho)$ and its first derivative, generating the field-temperature curves is then a matter of

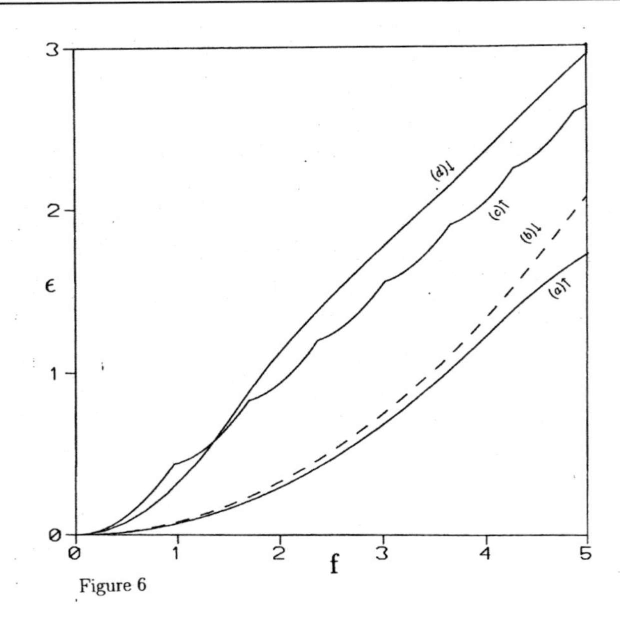

Figure 6: The critical field-temperature curves for some type II superconducting filaments: (a) solid cylinder in an azimuthal magnetic field, (b) the constant $|\psi|$ approximate result for the system of curve (a), (c) the result for a solid cylinder in a uniform parallel magnetic field and (d) the result for a thin film in a uniform parallel magnetic field; after Masale [21].

standard routine. In this case, the search is for $k_z R$, the so-called center of the 'cyclotron' orbit, which gives the lowest eigenvalue ε.

Figure 6 shows the temperature-field curves of some type II superconducting filaments as follows [21]: (a) the full numerical solutions of equation (24), (b) the corresponding constant ψ approximation generated from equation (25), (c) and (d) are the results for a solid cylinder and a thin film in a uniform parallel applied magnetic field, respectively. For the reason stated earlier about representing the field, here as f_s, the results shown here are directly comparable despite the spatial inhomogeneity of the azimuthal applied magnetic field. Nonetheless, a significant enhancement of the critical temperature is predicted for the azimuthal as opposed to the axial orientation of the magnetic field; compare curves (a) and (c) of figure 6. Note that within the range of the field-values used, there is very good agreement between the 'exact' and the constant ψ approximate results, curves (a) and (b) of figure 6, respectively. Since here k_z is a continuous variable, the presence of the azimuthal field, irrespective of its strength, gives rise to the existence of the $k_z > 0$ bound states. As a consequence of this, flux-entries occur for all non-zero magnetic field values. It is nonetheless interesting to note that for $k_z R = 0$, the $\epsilon - f$ curve of the system considered here follows closely, and almost overlaps the critical field-temperature curve of a thin film, at least for the range of the f values used. The eigenvalues corresponding to $k_z \leq 0$ are larger than those for $k_z > 0$ and therefore have no physical significance. Note also that for very fine filaments (small R or d) all the curves shown in figure 6 are parabolic in f. For large f (corresponding to a thick cylinder) the curvature of the cylinder can hardly matter,

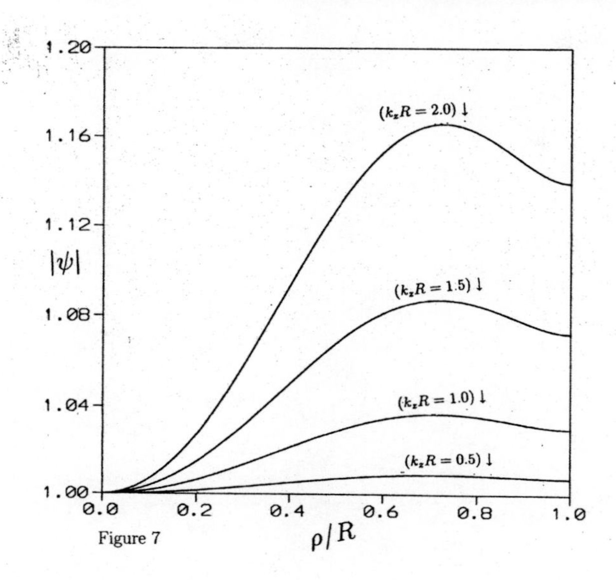

Figure 7: The wavefunction profiles across the cylinder for some points on the field-temperature curve shown as figure 6 (a). In particular, the fluxoid values, indicated against each of the corresponding curves are: $k_z R = 0.5$ for the curve, increasing in steps of 0.5, up to 2.0 for the highest curve; after Masale [21].

the result for a flat surface in a parallel magnetic field should then be recovered. This is not the case here and as such critical temperature of this system is significantly higher than that of a thin film. To this end some few observations are noted as follows: The induced magnetic field is not uniform but increases linearly with the radial distance. This implies that the innermost regions of a thick cylinder, where the field is less than H_{c2}, should be in the superconducting phase. The region just outside the radius R_{c2}, corresponding to the magnetic field contour value of H_{c2}, should then be in the normal state. A superconducting sheath, of course, persists at the surface of the cylinder for fields up to the higher parallel surface nucleation field H_{c3}. It is thought, loosely speaking, that this system then more or less behaves like an hollow cylinder of imaginary thickness $\sim [R - R_{c2}]$. This effect of the enhanced critical temperature is consistent with that of the reduced dimensionality in real systems.

Figure 7 shows the radial wavefunction profiles for some points on the azimuthal critical field-temperature curve of figure 6(a) [21]. Each curve corresponds to a different value of the fluxoid number. These are: $k_z R = 0.5$, for the lowest curve, increasing in steps of 0.5 up to 2.0 for the highest curve. For small values of f, that is, for $R < \xi$, the radial function can hardly change across the cylinder radius and the cylinder nucleates uniformly across its thickness. Phase changes for χ begin to emerge for intermediate values of the field, $R \sim \xi$. For $\xi > R$, there is a build up of the wavefunction towards the surface of the cylinder, which confirms the presence of a superconducting sheath near the surface. It is seen from figure 7 that for relatively large values of f the radial wavefunctions are characterized by a peak. The peak may be viewed as signifying regions of the cylinder where the superconducting

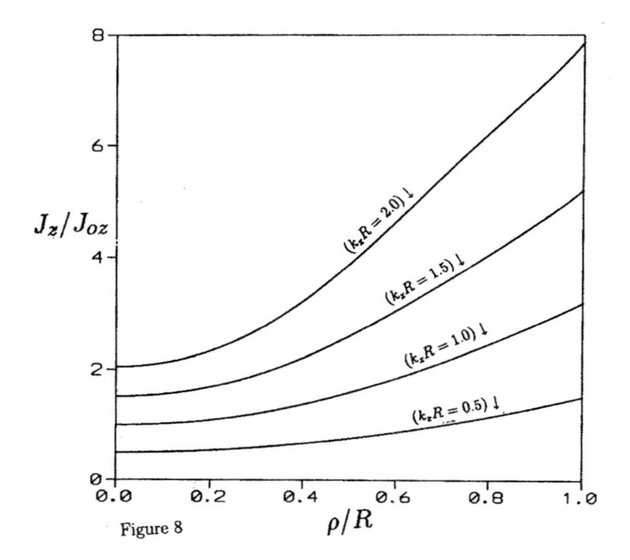

Figure 8: The distribution of the quantum mechanical current density across the cylinder radius corresponding to the wavefunctions shown in figure 7. The curves stack-up according to the increasing fluxoid numbers as: $k_z R = 0.5; (0.5); 2.0$; after Masale [21].

phase is most stable. Once again, note that the field at the radial distance corresponding to this peak is between H_{c2} and H_{c3}. Note that in contrast to the case of a cylinder in a uniform parallel magnetic field, in this case the radial wavefunctions for large f are finite at the origin. This implies the presence of the superconducting phase for all regions of the cylinder irrespective of its thickness. Further, considering the fine grid of the vertical axis, the radial function remains more or less constant across the cylinder thickness, at least for the values of f used. It is hardly surprising, therefore, that the 'exact' and the constant ψ approximate results are in very good agreement.
Since here the radial wavefunction is assumed to be real and only the z-component of the vector potential is considered, only the z-component J_z of the current density is non-zero and is given by

$$\frac{J_z}{J_{oz}} = \chi^2(\varrho)[|k_z|R + 2f\varrho] \qquad (26),$$

where $J_{oz} = 2e\hbar/(\mu R)$.
Figure 8 illustrates the functional spatial variation of the current densities corresponding exactly to the wavefunctions shown in figure 7 [21]. This variation of the current densities confirm the picture given by figure 7, in particular, that at least in the range of the f values used, all regions of the solid cylinder may be said to be in the superconducting phase.

4.2 Current-Carrying Core Surrounded by a Bulk Superconductor

A brief description of the idealized system considered is as follows: A current I of uniform density is thought to be passed along the axis of an infinitely long cylindrical core of radius R. The current-carrying core is then enveloped by a thick isotropic type II superconductor, in general, of length L_z. Ideally, the core could be fashioned out of a superconductor of a judiciously higher critical temperature than that of the surrounding material. This is to ensure that the surrounding superconductor reaches the field-induced transition to the normal state at lower fields than the core-material. Candidates for the core-material are any of high-temperature superconductors, including the extensively studied polycrystalline compound of magnesium diboride: MgB_2 [27,28]. In order to prevent current leakage, the core could be plated with a normal metal of very small thickness but large value of the G-L extrapolation length, δ. From a theoretical point of view, the limit $\delta \to \infty$, means that the boundary condition for a superconductor-insulator interface given by equation (4) holds well. The vector potential of the induced magnetic field is taken in the gauge:

$$A_z = -\frac{1}{2}B_s R[1 + 2\ln(\rho/R)]. \tag{27}$$

Note that the first term: $-\frac{1}{2}B_s R$, is included here to ensure continuity of the overall potential in crossing the boundary $\rho = R$ from the inner regions of the current-carrying core. Following exactly the same analysis outlined in section 3.0, solutions of the linearized G-L equation for the order parameter are found in terms of the confluent hypergeometric function. Note that the first of the approximations given by equation (10) leads to a form of the vector potential which corresponds to a uniform azimuthal magnetic field of magnitude: $B = c_2 B_s$. The application of the insulator-superconductor boundary condition leads to the following eigenvalue equation:

$$U(a, 1, \zeta_R) + 2aU(a + 1, 2, \zeta_R) = 0, \tag{28}$$

where

$$a = \frac{1}{2} + \frac{1}{2}\frac{\lambda_1}{\sqrt{\lambda_2}} \text{ and } \zeta_R = 2\sqrt{\lambda_2} f, \tag{29}$$

in which

$$\lambda_1 = k_z^2 R^2/4f - \varepsilon/f + [1+2c_1+d_1]f - [1+c_1]k_z R \text{ and } \lambda_2 = d_2 + 2c_2[1 - k_z R/2f]. \tag{30}$$

Again, since here A_z is negative, minimum contribution of the axial kinetic-energy-type to the Landau energy corresponds to zero or positive values of the axial wave number. Note that since the required eigenvalue corresponds to $k_z \geq 0$, both the parameter a and the argument ζ_R can switch between the real and complex planes. Based on a knowledge of the properties of the eigenfunctions, the lowest value of ε is obtained for very small argument of the function U, that is, $\zeta_R \to 0$. For finite values of the field, the vanishing of the argument corresponds to the limit: $\lambda_2 \to 0$. Now, one of the difficulties that arises in the numerical solutions of equation (28) is that the function U diverges rapidly as $\zeta_R \to 0$. A fairly good approximate form of ε, obtained on imposing the condition that $\lambda_1 \to 0$ as $\lambda_2 \to 0$, is given by

$$\varepsilon \approx \frac{1}{4}k_z^2 R^2 - [1 + c_1]|k_z|Rf + [1 + 2c_1 + d_1]f^2, \tag{31}$$

where k_z takes values such that $|k_z|R = [2 + d_2/c_2]f$. In fact this result may be obtained from equation (28), employing the appropriate limiting forms of the function U for small argument.

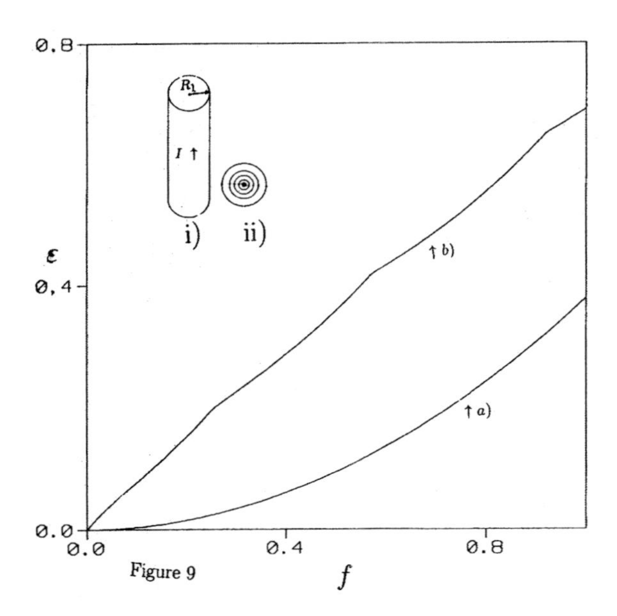

Figure 9: The critical field-temperature curves essentially of a cylindrical hole subjected to: a) a current-induced azimuthal magnetic field and b) a uniform parallel magnetic field. The insert depicts: i) a schematic diagram of the system modeled and ii) a sketch of the magnetic field lines corresponding to a current I directed out of the plane of the paper; after Masale [22].

Figure 9 shows the universal temperature-field curves essentially of a hole of radius R for two different orientations of the applied magnetic field: a) azimuthal and b) axial. The insert shows: i) the geometry of the problem considered and ii) a sketch of the magnetic field lines corresponding to a current I directed out of the plane of the paper. The results plotted in figure 9a) were computed exactly from equation (31). The constants for the linear fittings of the logarithmic terms are: $c_1 = 0.341$, $c_2 = 0.220$, $d_1 = -0.613$ and $d_2 = 0.616$, corresponding to the range $1 \leq \rho/R \leq 10$ of the radial distance. It is seen in figure 9 that, for a given temperature, the azimuthal critical field is much greater than its parallel analogue. This is precisely the prediction of an earlier study [21]. As mentioned earlier, verification of the results presented here should be fairly straightforward in the well established experimental techniques, for example, such as by Fomin *et al.* [18] or Strunk *et al.* [19]. Recall that the core is thought be fashioned out of a material of a judiciously higher critical temperature than that of the superconducting specimen. In view of this presumption and for no metallic cladding with a normal metal, one experimental procedure might be the following: while maintaining a constant current at a sufficiently high initial temperature, the sample is then cooled to or just below its critical temperature. This way, only the core-material should be in the superconducting state. The onset of the superconducting phase of the sample, at the critical temperature, should then be marked by a "leakage" of the current from the core into the sample and therefore a significant decrease of the current flowing through the core-material. The procedure outlined above can be carried out iteratively to accumulate the correspon ding pair of the critical parameters necessary for the construction of the universal temperature-field curve depicted in figure 9a).

4.3 Hollow Cylinder Cladding a Current-Carrying Core

The analysis of the previous section is now extended to the system of a type II superconducting hollow cylinder of length L_z and thickness $(R_2 - R_1)$ cladding the current-carrying core of radius R_1. The eigenvalue equation for this system is exactly that derived in section 3.0, of course, with appropriate replacements of the argument and the relevant parameters as follows:

$$a = \frac{1}{2} + \frac{1}{2}\frac{\Lambda_1}{\sqrt{\Lambda_2}}, \quad b = 1 \quad \text{and} \quad \zeta_{R_2} = 2\sqrt{\Lambda_2} f_2, \tag{32}$$

where

$$\Lambda_1 = k_z^2 R_2^2/4f_2 - \varepsilon_2/f_2 + [(1+2c_1)(1-\ln\eta) + d_1]f_2 - [1 + c_1 - \ln\eta]k_z R_2 \tag{33}$$

and

$$\Lambda_2 = d_2 + 2c_2[1 - \ln\eta - k_z R_2/2f_2] \tag{34}$$

in which $f_2 = \pi B_s R_2^2/2\phi_o$.

Following the same argument leading to equation (31) above, that is, imposing the condition that $\Lambda_1 \to 0$ as $\Lambda_2 \to 0$, a crude form of the critical temperature is found to be

$$\varepsilon_2 \approx \frac{1}{4}k_z^2 R_2^2 - [1 + c_1 - \ln\eta]|k_z|R_2 f_2 + [(1+2c_1)(1-\ln\eta) + d_1]f_2^2, \tag{35}$$

where k_z now takes values such that $|k_z|R_2 = [2(1-\ln\eta) + d_2/c_2]f_2$.

A more refined approximate limiting form of ε_2 is given by the so-called constant ψ approximate result which, after some cumbersome algebra, is found to be:

$$\varepsilon = \frac{1}{2}k_z^2 R_2^2 \frac{1}{1+\eta} - \frac{1}{2}s_1|k_z|R_2 f_2 + \frac{1}{2}s_2 f_2^2, \tag{36a}$$

where

$$s_1 = \frac{4}{1+\eta} - 1 - 2\Big[\frac{2}{1+\eta} + \frac{\eta^2}{1-\eta^2}\Big]\ln\eta \tag{36b}$$

and

$$s_2 = \frac{4}{1+\eta} + 1 - \Big[\frac{8}{1+\eta} + \frac{\eta}{1-\eta^2}\Big]\ln\eta + 2\Big[\frac{2}{1+\eta} - \frac{\eta^2}{1-\eta^2}\Big]\ln^2\eta. \tag{36c}$$

Now, in the case of a specimen of finite length L_z, the axial wave number is found to satisfy the following relationship: $\sin k_z L_z = 0$. In general, k_z is quantized according to: $k_z = n\pi/L_z;\ n = 0, \pm 1, \pm 2, \ldots$. Of course, for an infinitely long specimen k_z may be treated as a continuous variable. It is perhaps worth mentioning that surface nucleation of superconductivity should also occur at the end faces of a short cylindrical shell. Taking this surface nucleation into account should be fairly straightforward since the explicit form of the current-induced magnetic field depends only on the dimensions of the core and not of

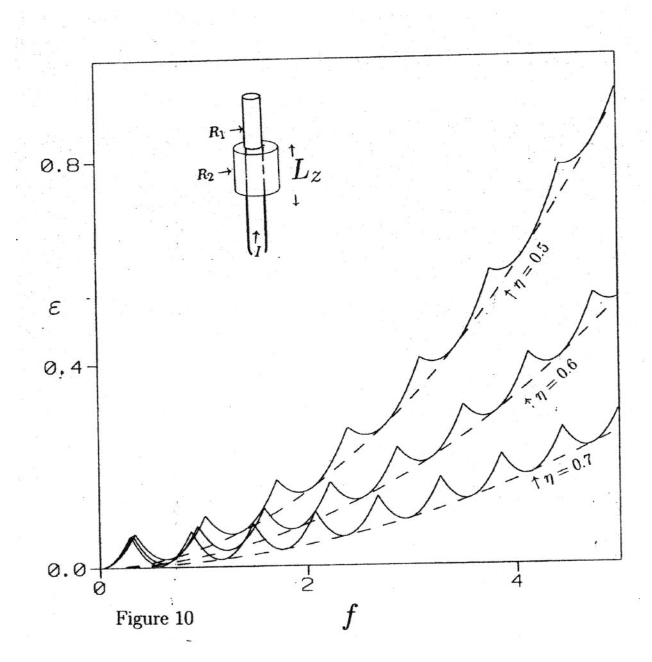

Figure 10: The two sets of $\varepsilon - f$ curves of a type II superconductor cladding a current-carrying core. The insert is a schematic geometry of the system studied; an infinitely long current-carrying core of radius R_1 tightly surrounded by a mesoscopic type II superconductor of radius R_2 and length L_z. Focusing on the dashed curves which are for an infinitely long specimen; $L_z \to \infty$; these stack-up according to the increasing values of the shell-thickness: $\eta = 0.7$ for the lowest curve, $\eta = 0.6$ for the middle curve and $\eta = 0.5$ for the highest curve. The smooth curves, characterized by quasi-periodic oscillations, are the results of a short specimen such that $|k_z|R_2 = \pi R_2/L_z = 1$. These also stack-up according to exactly the values of η given above for the broken curves; after Masale [22].

the specimen. The surface area of the specimen which can support nucleation of superconductivity there is very small. For this reason, this nucleation effect at the end-surfaces of the mesoscope is not considered.

Figure 10 shows two sets of the $\varepsilon_2 - f_2$ curves of mesoscopic systems which differ only in thickness. The insert is a schematic diagram of the system studied; basically an infinitely long current-carrying core of radius R_1 tightly surrounded by a mesoscopic type II superconductor of radius R_2 and length L_z. Focusing on the dashed curves, these stack-up according to the increasing values of the shell-thickness as follows: $\eta = 0.7$ for the lowest curve, $\eta = 0.6$ for the middle curve and $\eta = 0.5$ for the highest curve. To be more specific, the dashed curves are for an infinitely long specimen and the smooth curves with cusps are for a short specimen such that $|k_z|R_2 = \pi R_2/L_z = 1$, that is, for quantized axial wave number. As seen in figure 10, minima of the cusps are superimposed on the smooth rising curves of a specimen of the same thickness but infinite length. For small fields, the critical field corresponds to $k_z R_2 = 0$, however, as the field is increased to some critical value; B_{cr}, say, $k_z R_2$ begins to deviate from zero. The initial deviation of $k_z R_2$ from zero signifies the first flux-entry point at which, at least qualitatively, it becomes possible for a vortex loop to fit inside the mesoscope. It is hardly surprising that the $\varepsilon_2 - f_2$ curves for a quantum ring are characterized by rather distinct flux-entry points since k_z is then a discrete variable. It is seen in figure 10 that the critical field is significantly enhanced as a result of reducing the dimensions, in particular, the thickness, of the superconducting specimen.

Finally, although spatial variations of ψ and J are not shown here, the superconducting phase is expected to dominate over all regions of the superconductor for all fields lower than $H_{c3}(\varphi)$. Arguably, the form of the applied magnetic field tends to counteract spatial variations of the superconducting phases of the specimen. This counter effect of spatial variations of the superconducting phases should lead to the suppression of flux-entries. It is anticipated, therefore, that the order parameter will be fairly constant more or less throughout the specimen, hence signifying the presence of the superconducting state.

5 Conclusions

Calculations were presented of the critical fields of isotropic type II superconducting filaments which posses cylindrical symmetry based on the linearized Ginzburg-Landau equation for the order parameter. The spatially inhomogeneous magnetic field was considered in two distinct orientations of the relative to the interface(s) of the superconducting structure; either axial or azimuthal. Spatial distributions of the superconducting phase, especially in thick specimen, were inferred from the spatial variations of the wave functions and the corresponding current densities.

The profile of the parallel applied magnetic field consisted of a logarithmic variation in the radial distance, this being superimposed on a constant background value. In the parallel configuration, particularly for near uniform applied magnetic field, the universal temperature-field curves were found to be characterized by a succession of distinct flux-entry points at each of which m decreases by unity. Much prominent cusps, known as Little-Parks oscillations, are a characteristic a very thin-walled annular disk. The quasi-periodicity of these cusps in the field variable (f) were found to increase substantially for stronger spatial variations of the applied magnetic field. In a mesoscopic structure, this was also accompanied by dramatic reduction of the amplitude of Little-Parks oscillations.

The azimuthal applied magnetic field was thought of as essentially that induced by a current of uniform density passed along the axis of an infinitely long core. Since the magnetic field lines are parallel to the superconductor-core heterointerface, the required critical field is in fact a direct analogue of the parallel critical field, H_{c3}. The results of these investigations point towards a substantial enhancement of the critical field when the magnetic field is applied in the azimuthal rather than in the axial direction. Potential applications of the results of these investigations may be found in, for example, employing superconducting electromagnets in the generation of very high magnetic fields.

In general, nucleation of superconductivity was found to be fairly uniform across the superconducting specimen for a larger range of the field-values compared to the case of a uniform applied magnetic field. This is what may be anticipated since the applied magnetic field attains its largest at the surface of the superconductor, where nucleation of superconductivity

is favourable, anyway. The profile of the applied magnetic field, axial or azimuthal, is inherently such as to counter surface nucleation. This has the effect, therefore, of quenching spatial distributions of the superconducting phase across the specimen. The gain in theoretical studies whereby profiles of the magnetic field are tailored in this manner is that the system can treated in terms of a *constant* wave function.
Finally, no specific conclusions can be drawn from these results about high-T_C superconductors since the theoretical treatment given is limited only to isotropic materials. However, in view of future potential applications in circuitry devices, it is worth extending these investigations, both theoretically and experimentally, to high-T_C materials.

References

[1] W. Meissner and R. Ochsenfeld, *Naturwissenschaften* **21** (1933) 787

[2] S. Gygax and R.H. Kropschot, *Phys. Lett.* **9** (1964) 91

[3] D. Saint-James and P.G. de Gennes, *Phys. Lett.* **7** (1963) 306

[4] E. Guyon, F. Meunier and R.S. Thompson, *Phys. Rev.* **156** (1967) 452

[5] S. Jin and J.E. Graebner, *Mater. Sci. Eng.* **B 7** (1991) 243

[6] M. Masale, N.C. Constantinou and D.R. Tilley, *Supercond. Sci. Technol.* **6** (1993) 287

[7] S. Takács, *Czech J. Phys.* **19** (1969) 1366

[8] N.C Constantinou, M. Masale and D.R. Tilley, *J. Phys C.: Condens. Matter* **4** (1992) L293

[9] W. A Little and R.D Parks, *Phys. Rev. Lett.* **9** (1962) 9

[10] M.V. Milošević, S.V Yampolskii and F.M. Peeters, *Phys. Rev. B* **66** (2002) 024515.

[11] M. Masale, *Physica C* **397** (2003) 29

[12] J. Reijniers and F.M. Peeters, *J. Phys.: Condens. Matter* **12** (2000) 9771

[13] H.-S. Sim, K.J. Chang, N. Kim and G. Ihm, *Phys. Rev.* **B 63** (2001) 125329

[14] M. Masale, *Phys. Scr.* **65** (2002) 273.

[15] M. Masale, *Phys. Scr.* **67** (2003) 136

[16] H.J. Fink and A.G. Presson, *Phys. Rev.* **151** (1966) 219

[17] S.V Yampolskii and F.M. Peeters, *Phys. Rev. B* **62** (2000) 9663.

[18] V. M. Fomin, V.R. Misko, J.T. Devreese and V.V. Moshchalkov, *Physica B* **249-251** (1998) 476

[19] C. Strunk, V. Bruyndoncx, C. Van Haesendonck, V.V Moshchalkov, Y. Bruynseraede, C.-J. Chien, B. Burk and V. Chandrasekhar, *Phys. Rev.* **B 57** (1998) 10 854

[20] A.K. Geim, I.V. Grigorieva, J.G.S. Lok, J.C. Maan, S.V. Dubonos, X.Q. Li, F.M. Peeters and Yu.V Nazarov, *Superlatt. Microstruct.* **23** (1998) 151

[21] M. Masale, *Physica C* **377** (2002) 75

[22] M. Masale, *Supercond. Sci. Technol.* **17** (2004) 993

[23] P.G de Gennes (1966) *Superconductivity of Metals and Alloys* (New York: Benjamin)

[24] D.R. Tilley and J. Tilley (1990) *Superfluidity and Superconductivity* (Bristol: Hilger)

[25] H.J. Fink and V. Grünfeld, *Phys. Rev. B* **22** (1980) 2289

[26] Robert Haslinger and Robert Joynt, *Phys. Rev.* **B 61** (2000) 4206

[27] İ. N. Askerzade, A. Gencer and Güçlü, *Supercond. Sci. Technol.* **15** (2002) L13

[28] R.H.T. Wilke, S.L. Bud'ko, P.C. Canfield, D.K. Finnemore, Raymond J. Suplinskas and S.T. Hannahs, *Physica C* **424** 1

In: Superconductivity, Magnetism and Magnets
Editor: Lannie K. Tran pp. 55-96
ISBN 1-59454-845-5

Chapter 3

Combined Effects of Disorder and Magnetic Field in Superconductors

P.D. Sacramento and J. Lages
[1]Centro de Física das Interacções Fundamentais,
Instituto Superior Técnico, Av. Rovisco Pais, 1049-001 Lisboa, Portugal
[2]Laboratoire de Physique Théorique, UMR 5152 du CNRS,
Université Paul Sabatier, 31062 Toulouse Cedex 4, France

Abstract

We consider the effects of disorder due to impurities acting as scatterers for the electrons and as pinning centers for vortices, induced by an external magnetic field, in a strongly type-II superconductor. We consider the cases of s-wave and d-wave pairing symmetries. The effect of disorder reflects in the density of states, particularly at low energies. Considering first that no impurities are present but that the vortices are pinned randomly, the density of states as a function of magnetic field increases at low energies, closing the gap in the s-wave case and becoming finite in the d-wave case. Including the effect of the scattering from impurities, placed randomly using the binary alloy model, the density of states and the local density of states (LDOS) have to be solved self-consistently. The results for the LDOS agree qualitatively with experimental results if most vortices are pinned to the impurities. Also, we consider the effect of impurities on the vortex charge.

1 Introduction

The interplay between superconductivity and a magnetic field has attracted interest for a long time. For sufficiently strong fields the Meissner phase is destroyed and a mixed state appears in the form of a quantized vortex lattice [1]. The superconductor order parameter has zeros at the vortex locations, through which the external magnetic field penetrates in the sample. Contrarily to previous understanding, the increase of the magnetic field intensity, and its associated diamagnetic pair breaking, is counteracted at very high magnetic fields by the Landau level structure of the electrons [2]. This leads to interesting properties in strongly type-II superconductors such as enhancement of the superconducting transition temperature at very high magnetic fields, where the electrons are confined to the lowest

Landau level [2]. Associated with the zeros of the order parameter in real space are gapless points in the magnetic Brillouin zone [3], which lead to qualitatively different behavior at low temperatures and high magnetic fields for s-wave superconductors [4]. These nodes are associated with the center of mass coordinates of the Cooper pairs and not to some internal structure like in d-wave superconductors. Lowering the magnetic field, a quantum level-crossing transition leads to a gapped regime for s-wave superconductors and to states localized in the vortex cores [5, 6], as obtained long ago for an isolated vortex. Below this transition the behavior of the mixed state is qualitatively different and extends down to the vicinity of the Meissner phase. In this regime the behavior of a s-wave superconductor is very different from a d-wave superconductor. This is the regime most easilly accessible experimentally since the observation of the high-field behavior is in general difficult, as explained later. The presence of the vortices in the system affects the quasiparticles in a non-trivial way, as we will see ahead.

On the other hand, the effect of disorder on superconductivity has also attracted interest for a long time. In the case of non-magnetic impurities and s-wave pairing, Anderson's theorem states that, at least for low concentrations, they have little effect since the impurities are not pair-breaking [7]. In d-wave superconductors however, non-magnetic impurities cause a strong pair breaking effect [8]. In the limit of strong scattering it was found that the lowest energy quasiparticles become localized below the mobility gap, even in a regime where the single-electron wave-functions are still extended [9]. This result has been confirmed solving the Bogoliubov-de Gennes equations with a finite concentration of non-magnetic impurities [10]. However, allowing for angular dependent impurity scattering potentials it has been found that the scattering processes close to the gap nodes may give rise to extended gapless regions [11]. The case of magnetic impurities in the s-wave case also leads to gapless superconductivity [12].

Several conflicting predictions have appeared in the literature regarding the effect of the presence of impurities. Some progress towards understanding the disparity of theoretical results has been achieved realising that the details of the type of disorder affect significantly the density of states [13]. Particularly in the case of d-wave superconductors, in contrast to conventional gapped s-wave superconductors, the presence of gapless nodes is expected to affect the transport properties. Using a field theoretic description and linearizing the spectrum around the four Dirac-like nodes it has been suggested that the system is critical. It was obtained that the density of states is of the type $\rho(\epsilon) \sim |\epsilon|^{\alpha}$, where α is a non-universal exponent dependent on the disorder, and that the low energy modes are extended states (critical metal) [14]. Taking into account the effects of inter-nodal scattering (hard-scattering) it has been shown that an insulating state is obtained instead, where the density of states still vanishes at low energy but with an exponent $\alpha = 1$ independent of disorder [15]. The addition of time-reversal breaking creates two new classes designated spin quantum Hall effect I and II, due to their similarities to the usual quantum Hall effect, corresponding to the hard and soft scattering cases, respectively [13]. The proposed formation of a pairing with a symmetry of the type $d + id$ breaks time-inversion symmetry [16] but up to now remains a theoretical possibility. On the other hand applying an external magnetic field naturally breaks time-reversal invariance and therefore it is important to study the density of states in this case.

In general, disorder is due to the presence of impurities which may either scatter the quasiparticles and/or may serve as pinning centers for the field induced vortices. The density of states of a dirty but homogeneous s-wave superconductor in a high magnetic field, where the quasiparticles scatter off scalar impurities, was considered using a Landau level basis [17]. For small amounts of disorder it was found that $\rho(\epsilon) \sim \epsilon^2$ but when the disorder is higher than some critical value a finite density of states is created at the Fermi surface. In the same regime of high magnetic fields, but with randomly pinned vortices and no impurities, the density of states at low energies increases significantly with respect to the lattice case suggesting a finite value at zero energy [18]. Refs. [19] and [20] considered the effects of random and statistically independent scalar and vector potentials on d-wave quasiparticles and it was predicted [20] that at low energies $\rho(\epsilon) \sim \rho_0 + a\epsilon^2$, where $\rho_0 \sim B^{1/2}$. The effect of randomly pinned *discrete* vortices on the spectrum of a d-wave superconductor was also considered recently [21].

The addition of impurities in zero magnetic field has been studied using the Bogoliubov-de Gennes (BdG) equations. It was found that the d-wave superconductivity is mainly destroyed locally near a strong scatterer. The superfluid density is strongly suppressed near the impurities but only mildly affected elsewhere [22]. No evidence for localization of the low energy states was found. The superfluid density is indeed suppressed but less than expected [10, 23] and, accordingly, the decrease of the critical temperature with disorder is much slower than previously expected, in accordance with experiments [24]. Similar results of an inhomogeneous order parameter were also obtained for s-wave superconductors [25].

A question we address here is the influence of the positional disorder of the vortices on the quasiparticle states of either a s- or a d-wave superconductor in an external magnetic field not considering, at first, the scattering off impurities. Typically, in a BCS-type superconductor, vortices and quasiparticles see different disorders, due to their different scales. Vortices are large objects and their cores, even in High Temperature Superconductors, are significantly pinned only by very strong disorder which is often quite localized – a fine example is a hole through a lattice of YBCO or BSCCO irradiated by ions and filled with such "columnar pins". In contrast, no low energy quasiparticles go through these columnar pins but are rather scattered by more microscopic forms of disorder.

Clearly the effect of the impurities on the quasiparticles also has to be taken into account. Therefore we also consider the combined effect of the impurities and the vortices on the low temperature properties. We conclude that the main effect is due to the vortex scattering. We review aspects of the effects of a magnetic field in superconductors in Section 2. In section 3 we focus on recent work of the combined effects. In this chapter we review different aspects of the field but the main focus is on topics the authors have worked on.

2 Effects of an External Magnetic Field

2.1 Meissner Effect and Mixed Phase

A particle of charge q and internal spin $\vec{S}$, in the presence of an external magnetic field, has its energy changed according to the substitution $\vec{p} \rightarrow \vec{p} - \frac{q}{c}\vec{A}$, where $\vec{A}$ is the vector potential, and due to the Zeemann term $-\vec{\mu} \cdot \vec{B}$, where $\vec{\mu}$ is the spin magnetic moment. In the simplest case, the Cooper pairs are formed of two electrons with opposite spins and opposite momenta. The Zeemann term tends to align the spins and the Lorentz force tends to move the two electrons away from each other, and therefore, in general, the effect of the magnetic field is to break the pairs. The effect of the diamagnetic term is to give a finite momentum of one electron with respect to the other in the pair, as a consequence of the broken time-reversal invariance. Moreover, in the superconducting state the photons acquire a mass through the Anderson-Higgs mechanism, leading to the Meissner phase where the magnetic field is expelled from the interior of the superconductor (completely in type-I superconductors). The shielding of the external field is caused by the appearance of surface currents that cancel perfectly the external field beyond a penetration depth, λ. Another length scale which is important is the coherence length, ξ, which measures the distance of the establishment of the superconducting order parameter into the superconducting region. The ratio of the two lengths, $k = \frac{\lambda}{\xi}$, determines different regimes of the superconducting materials. For small k ($< 1/\sqrt{2}$) the energy of the interface between the normal phase and the superconducting phase is positive (type-I superconductors) and the magnetic field is expelled from the interior of the superconductor (perfect diamagnet). The field lines are bent close to the interface and this may be favorable energetically depending on the shape of the superconducting material. However, if k is large ($> 1/\sqrt{2}$) the surface energy is negative and it is favorable that the total area of contact between the two phases is large. In this case (type-II superconductors) the field penetrates into the superconducting region in flux tubes surrounded by superconducting regions (mixed phase). Due to the phase coherence of the superconductor order parameter, these flux lines are quantized in units of the flux quantum $\Phi_0 = \frac{hc}{2e}$. As the magnetic field density increases, the density of the flux lines also increases until their cores overlap and the system becomes non-superconducting.

At low fields the magnetic field is totally expelled, except for a surface slab of thickness λ, and the system is in the Meissner phase. As the field increases the system allows the penetration of flux tubes, starting at H_{c1}, and enters the mixed phase. If the system is strongly of type-II, $k >> 1$, the field H_{c2}, where the system enters the normal phase, is such that $H_{c2} >> H_{c1}$. In the mixed phase there are supercurrents around each vortex that interact with the quasi-particles in a non-trivial way.

For fields $H_{c1} < H << H_{c2}$ the behavior of the system is dominated by the scattering of the quasiparticles off the vortices and the supercurrents. The quasiparticle spectrum retains a character that at low energies is a result of the combined effect of the bound states in the vortex cores and states in the continuum, due to the propagation in the superconducting region (between the vortex cores).

In the high field limit, close to H_{c2}, since the critical field is very large if $k >> 1$, the nature of the spectrum is qualitatively different. At these very high fields and since λ is large, the magnetic field is almost uniform throughout the system. Close to the transition to the normal phase, the Ginzburg-Landau approach can be used. It was concluded that the vortices in this regime are organized most favorably in a triangular lattice (Abrikosov lattice) with a lattice constant which is proportional to the magnetic length, $l = \sqrt{\frac{\hbar c}{eH}}$, defined as $\Phi/\Phi_0 = S/\pi l^2 = N_\Phi$, where N_Φ is the number of flux quanta in the area S (perpendicular to the magnetic field). The vortex cores are not far apart and the decay of the field in the intermediate space region is negligible. Therefore, we have a system of charges in the presence of an almost uniform external field (the screening is negligible). The states of the system are therefore expected to be of an extended nature, and the vector potential is of the type $A \sim Br$, where r is the space coordinate. Therefore, for a large system A is large and, since the Hamiltonian depends on the magnetic field through the vector potential, the effect of the field can not be treated in perturbation theory. This leads naturally to the use of the Landau levels as an appropriate basis in this regime. It was concluded that the quasiparticle spectrum is gapless even for a s-wave superconductor due to the coherent nature of the quasiparticles. This property has important consequences as we shall recall later.

In this work we are interested in looking at various regimes of the phase diagram for both s- and d-wave symmetries and to study the effects of disorder.

2.2 Bogoliubov-de Gennes Equations

A convenient way to study a superconductor in the presence of an arbitrary magnetic field and in the presence of an arbitrary distribution of impurities is through the Bogoliubov-de Gennes equations (BdG) [26]. These are equations for the amplitudes $u(\vec{r})$ and $v(\vec{r})$ defined through the transformation from the field operators for the electrons in point $\vec{r}$ and spin $\alpha = \uparrow, \downarrow$, $(\psi(\vec{r}, \alpha), \psi^\dagger(\vec{r}, \alpha))$ to the quasiparticle operators $(\gamma_n, \gamma_n^\dagger)$

$$\begin{aligned} \psi(\vec{r}, \uparrow) &= \sum_n \left(\gamma_{n,\uparrow} u_n(\vec{r}) - \gamma_{n,\downarrow}^\dagger v_n^*(\vec{r}) \right) \\ \psi(\vec{r}, \downarrow) &= \sum_n \left(\gamma_{n,\downarrow} u_n(\vec{r}) + \gamma_{n,\uparrow}^\dagger v_n^*(\vec{r}) \right) \end{aligned} \tag{1}$$

that diagonalize the mean-field (BCS) Hamiltonian as

$$H_{eff} = E_g + \sum_{n,\alpha} \epsilon_n \gamma_{n,\alpha}^\dagger \gamma_{n,\alpha} \tag{2}$$

where E_g is the groundstate energy (no excitations) and the energies ϵ_n are obtained from the solution of the BdG equations. In the s-wave case these can be written as

$$\begin{aligned} (H_e + U(\vec{r}))\, u(\vec{r}) + \Delta(\vec{r}) v(\vec{r}) &= \epsilon u(\vec{r}) \\ -\left(H_e^* + U(\vec{r})\right) v(\vec{r}) + \Delta^*(\vec{r}) u(\vec{r}) &= \epsilon v(\vec{r}) \end{aligned} \tag{3}$$

Here

$$H_e(\vec{r}) = \frac{1}{2m} \left(\frac{\hbar}{i} \vec{\nabla} - \frac{e}{c} \vec{A} \right)^2 + U_0(\vec{r}) - E_F$$

where m is the electron mass, $\vec{A}$ is the vector potential, $U_0(\vec{r})$ an external potential, E_F the Fermi energy and $U(\vec{r})$ is the potential originated from the interactions between the electrons. The pairing is to be determined self-consistently together with the interaction potential

$$\begin{aligned} \Delta(\vec{r}) &= -V < \psi(\vec{r}, \downarrow) \psi(\vec{r}, \uparrow) >= V < \psi(\vec{r}, \uparrow) \psi(\vec{r}, \downarrow) > \\ U(\vec{r}) &= -V < \psi^\dagger(\vec{r}, \uparrow) \psi(\vec{r}, \uparrow) >= -V < \psi^\dagger(\vec{r}, \downarrow) \psi(\vec{r}, \downarrow) > \end{aligned} \tag{4}$$

where V is the coupling between the electrons, and we are considering spin singlet pairing.

For the case of a d-wave superconductor the BdG equations can be written as

$$\begin{aligned} \left(H_e + U(\vec{R})\right) u(\vec{R}) + \int d\vec{r} \Delta(\vec{R} - \frac{\vec{r}}{2}, \vec{r}) v(\vec{R} - \vec{r}) &= \epsilon u(\vec{R}) \\ -\left(H_e^* + U(\vec{R})\right) v(\vec{R}) + \int d\vec{r} \Delta^*(\vec{R} - \frac{\vec{r}}{2}, \vec{r}) u(\vec{R} - \vec{r}) &= \epsilon v(\vec{R}) \end{aligned} \quad (5)$$

where $\vec{R}$ and $\vec{r}$ are the center of mass and relative coordinates, respectively, and

$$\Delta(\vec{R}, \vec{r}) = V(\vec{r}) < \psi(\vec{R} - \frac{\vec{r}}{2}, \uparrow) \psi(\vec{R} + \frac{\vec{r}}{2}, \downarrow) > \quad (6)$$

2.3 Vortex States

We begin by considering a single vortex in a superconductor.

2.3.1 Caroli, de Gennes and Matricon States

The nature of the low energy states in the presence of a single vortex with s-wave symmetry was solved long ago [6]. There are localized bound states in the vortex core in addition to states in the continuum. The lowest energy states for a s-wave vortex were obtained analytically by Caroli, de Gennes and Matricon (CGM) [6]. The energy of the lowest eigenvalues of the form

$$E_n \sim \frac{\mu \Delta^2}{E_F} \quad (7)$$

These are localized states at the vortex core. Moreover, they obtained that the density of states is $N(\epsilon) \sim N(0)\xi^2$, with the interpretation that the vortex line is equivalent to a normal region of radius ξ. Since the low lying states are bound to the vortex, the specific heat should be linear in T, as in a normal conductor.

The general solution is obtained solving numerically the BdG equations. Eq.(3) is solved using the decompositions [27]:

$$\begin{aligned} u^i(\rho, \varphi) &= \sum_{\mu, j} c^i_{\mu, j} e^{i\varphi(\mu - \frac{1}{2})} \phi_{j, \mu - \frac{1}{2}}(\rho) \\ v^i(\rho, \varphi) &= \sum_{\mu, j} d^i_{\mu, j} e^{i\varphi(\mu + \frac{1}{2})} \phi_{j, \mu + \frac{1}{2}}(\rho) \end{aligned} \quad (8)$$

and choosing a gauge such that $\Delta(\rho, \varphi) = \Delta(\rho) e^{-i\varphi}$. Here $\mu = \pm\frac{1}{2}, \pm\frac{3}{2}, \pm\frac{5}{2}, \cdots$ is the planar angular momentum, j is the number of the zero of the Bessel functions in a disc of radius R and the normalized functions $\phi_{j,m}$ constitute a complete set over the zeros and are defined by

$$\phi_{j,m} = \frac{\sqrt{2}}{R J_{m+1}(\alpha_{j,m})} J_m \left(\alpha_{j,m} \frac{\rho}{R} \right) \quad (9)$$

Here J_m is the Bessel function of order m and $\alpha_{j,m}$ is the jth zero of the Bessel function J_m. The functions $\phi_{j,m}$ by construction are zero at the border of the disc $\rho = R$. In the basis of the functions $\phi_{j,m}$ we have to solve a set of equations for the c and d coefficients of the form

$$\sum_{\mu',j'} \begin{pmatrix} T^{-}_{\mu,j;\mu',j'} & \Delta_{\mu,j;\mu',j'} \\ \Delta^{T}_{\mu,j;\mu',j'} & T^{+}_{\mu,j;\mu',j'} \end{pmatrix} \begin{pmatrix} c_{\mu',j'} \\ d_{\mu',j'} \end{pmatrix} = E \begin{pmatrix} c_{\mu,j} \\ d_{\mu,j} \end{pmatrix}$$

Here the various components are diagonal in the angular momentum $T^{\pm}_{\mu,j;\mu',j'} = \delta_{\mu,\mu'} T^{\pm}_{\mu,;j,j'}$ and the same for the off-diagonal terms. The various terms are given by

$$\begin{aligned} T^{\pm}_{\mu;j,j'} &= \mp \left(\frac{\hbar^2}{2m} \left(\frac{\alpha_{j,\mu\pm 1/2}}{R} \right)^2 - E_F \right) \delta_{j,j'} \\ &\mp \frac{1}{2m} \int_0^R d\rho\rho\phi_{j,\mu\pm1/2}(\rho) \left(\frac{e}{c}\right)^2 A^2_{\varphi}(\rho)\phi_{j',\mu\pm1/2}(\rho) \\ &- \frac{1}{2m} \int_0^R d\rho\rho\phi_{j,\mu\pm1/2}(\rho) \frac{2e}{c}\hbar(\mu \pm 1/2)\frac{A_{\varphi}(\rho)}{\rho}\phi_{j',\mu\pm1/2}(\rho) \end{aligned}$$

and

$$\Delta_{\mu;j,j'} = \int_0^R d\rho\rho\phi_{j,\mu-1/2}(\rho)\Delta(\rho)\phi_{j',\mu+1/2}(\rho) \tag{10}$$

The physical quantities are obtained in the usual way

$$\Delta(\rho) = V \sum_{\mu,i(|E_i|\leq\omega_D)} \bar{u}_{\mu,i}(\rho)\bar{v}_{\mu,i}(\rho)\left(1 - 2f(E_{\mu,i})\right) \tag{11}$$

with $u_{\mu,i}(\rho,\varphi) = e^{i\varphi(\mu-1/2)}\bar{u}_{\mu,i}(\rho)$ and $v_{\mu,i}(\rho,\varphi) = e^{i\varphi(\mu+1/2)}\bar{v}_{\mu,i}(\rho)$. Also, the current is given by

$$\begin{aligned} j_{\varphi}(\rho) &= -\frac{2|e|\hbar}{m} Re \sum_{\mu,i} (f(E_{\mu,i})\bar{u}_{\mu,i}(\rho)^2 \left((\mu - 1/2)\frac{1}{\rho} + \frac{|e|}{\hbar c} A_{\varphi}(\rho) \right) \\ &+ (1 - f(E_{\mu,i}))\bar{v}_{\mu,i}(\rho)^2 \left(-(\mu + 1/2)\frac{1}{\rho} + \frac{|e|}{\hbar c} A_{\varphi}(\rho) \right)) \end{aligned}$$

and the magnetic field and the vector potential. are obtained solving the Maxwell equation

$$\vec{\nabla} \times \vec{B} = \frac{4\pi}{c}\vec{J} = -\vec{\nabla}^2 \vec{A} \tag{12}$$

The BdG equations are then solved self-consistently. The energy levels show bound states as predicted by Caroli, de Gennes and Matricon together with states in the continuum. Both the coherence length and the penetration length are obtained self-consistently.

At low temperatures $T < T_c/k_F\xi$, where k_F is the Fermi momentum and ξ the coherence length, the quantum limit is reached and the various physical quantities show Friedel-like oscillations [28] with wave length $\sim 1/k_F$. The eigenfunctions also display the same oscillatory behavior. As the temperature lowers, the size of the vortex shrinks (Kramer and Pesch effect), and decreases until a minimum value is reached. The lowest eigenvalues corresponding to the bound states of CGM are not particle-hole symmetric, as evidenced by the Local Density of States (LDOS).

The d-wave symmetry case led to some early controversy but it was eventually clearly demonstrated that the low energy states, even though strongly peaked near the vortex core, extend along the four nodal directions and are indeed delocalized, as shown by the behavior of the inverse participation ratio (IPR) [29].

The BdG equations for a d-wave vortex can be solved expanding

$$\Delta(\vec{R},\vec{r}) = \Delta(R,\theta;\varphi) = \sum_{p,l} e^{-ip\theta} e^{il\varphi} \Delta_{pl}(R) \tag{13}$$

where $\vec{R} = (R,\theta)$ is the center of mass coordinate and $\vec{r}$ is the relative coordinate, as before. The integer p characterizes the winding of the superconducting phase around the vortex and l characterizes the orbital state of the pair. The wave functions are expanded in the Bessel functions as for the s-wave case like

$$\begin{aligned} u(\vec{R}) &= \sum_{\mu,m} e^{i\mu\theta} \Phi_{\mu m}(R) u_{\mu m} \\ v(\vec{R}) &= \sum_{\mu,m} e^{i\mu\theta} \Phi_{\mu m}(R) v_{\mu m} \end{aligned} \tag{14}$$

where μ is an integer and, as before,

$$\Phi_{\mu m}(R) = \frac{\sqrt{2}}{R_0 J_{\mu+1}(\alpha_{\mu m})} J_\mu\left(\alpha_{\mu m}\frac{R}{R_0}\right) \tag{15}$$

having considered a disc of radius R_0,μ is the angular momentum and $\alpha_{\mu m}$ is the mth zero of J_μ. In this basis the BdG equations are converted once again into an eigenvalue problem for the coefficients $u_{\mu m}$ and $v_{\mu m}$. In contrast to the s-wave case, where the angular momentum is a good quantum number and the diagonalization factorizes for each angular momentum value, in the d-wave case the various coefficients are coupled and one has to solve a full 2d problem [29].

The solution leads to the absence of bound states in the core and all states are extended, corresponding to the gap nodes due to the internal symmetry of the Cooper pairs. Consequently, the spectrum is continuous and gapless.

In the d-wave case other components of the gap function with other symmetries are induced in the vortex core. Far from the core region the symmetry is d-wave but close to the vortex s-wave order can be induced and the two symmetries coexist [30].

2.3.2 Vortex Charge

Some time ago [31] it was proposed that the vortex induced by an external magnetic field in a type-II superconductor should be electrically charged. This effect was proposed to occur since the chemical potential is expected to be larger in the vortex core than in the bulk of the superconductor. It is energetically favorable for the electrons to lower their energy through the condensation energy and, since the vortex core is interpreted as being a normal region, the electrons tend to move to the bulk leaving a charge deficiency close to the vortex line. Soon after it was proposed that the effect could be tested experimentally due to the dipole field created at the surface of the superconductor [32]. It has been claimed that the charge of the vortex has been measured in high temperature superconductors using NMR [33].

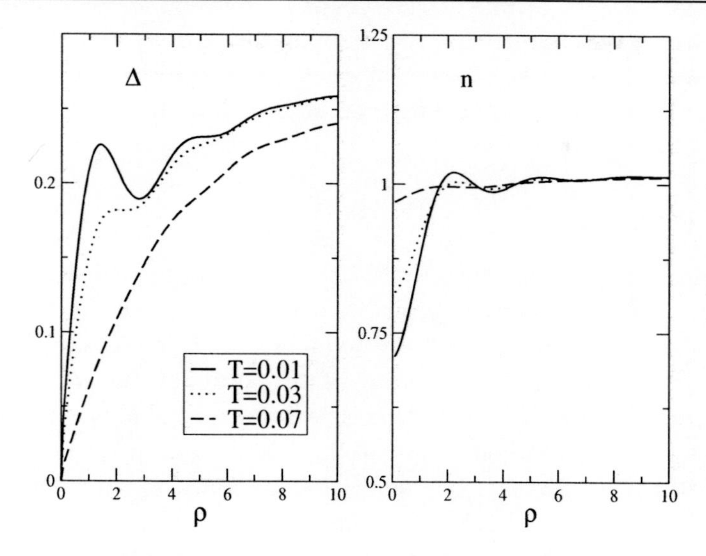

Figure 1: Δ and electron density as a function of distance for different values of T.

Theoretical studies on the existence of the vortex charge were carried out subsequently [34, 35, 36, 37]. In particular, a relation was established between the vortex core charge and the vortex bound states for an s-wave vortex [34]. However, in d-wave superconductors the low lying states are not localized [29]; also, in the high magnetic field regime, where a Landau level description is appropriate, the low lying states are coherent through the vortex lattice [4] both in the s-wave [4] and in the d-wave cases [38]. To our knowledge in this regime the vortex charge has not been studied.

Taking into account the screening of the vortex charge, the Friedel oscillation in the charge profile was obtained for the case of a s-wave vortex [35], showing that the charge is screened but prevails with a somewhat reduced value. The cases of other pairing symmetries were also considered [36, 39] showing that the existence of the vortex charge is universal. In Fig. 1 we present in the left panel Δ as a function of distance at a low temperature. In the right panel we show the electron density for the same temperature values. The oscillations of both quantities are clearly seen. The effect of temperature is to decrease the vortex charge. Also the size of the normal core region increases with temperature (Kramer-Pesch effect).

In all these studies the vortex is charged positively (electron defficiency). This positive charge has been argued to be the cause of the Hall anomaly, where the Hall conductance changes sign when entering the superconducting phase [40]. However, it has been shown that in the d-wave superconductors it is likely that other orderings compete with the superconductivity. In particular, antiferromagnetism [41], or d-density waves [42]. The vortex structure, and in particular the vortex charge, have been studied when there is competition between various order parameters [43, 39, 44, 45, 46]. Since these other order parameters compete with superconductivity, and this is absent in the vortex cores, implies that, in general a small region around the vortex will have a non-vanishing order parameter, which affects the density of states. Also, the vortex charge may change sign from an electron defficiency to an electron abundance at the vortex core. Increasing the temperature in a regime of parameters where the competing order is absent the positive vortex charge is recovered [44]. Later on we will study the effect of disorder on the vortex charge ignoring competing orderings. The effect of these competing orderings on the density of states will be referred ahead.

2.4 High Field Regime

Observation of the de Haas-van Alphen (dHvA) oscillations in many extreme type-II superconducting materials [47] clearly indicate the presence of Landau level (LL) quantization within a high-magnetic field (H), low-temperature (T) "pocket", surrounding the $H_{c2}(0)$ point in the $H-T$ phase diagram. The size of this pocket in conventional type-II systems (like Nb) is expected to be negligible. This is so because, within the BCS theory, the scale of the cyclotron splitting between LLs near $H_{c2}(0)$, $\omega_c \sim \omega_{c2}(0) \equiv eH_{c2}(0)/m^*c$, is set by the condensation energy, $\sim T_{c0}^2/E_F$, and should be much smaller than either the thermal smearing, $\sim T$, or the gap $\Delta(T,H)$, the scale for both being T_{c0}. Additional smearing due to disorder, Γ, makes this tiny pocket effectively irrelevant. In clean, intrinsically extreme type-II systems, the situation is significantly different. In extreme type-II systems $\omega_{c2}(0)$ is *comparable* to T_{c0} and there is large pocket in the phase diagram in which the LL structure *within* the superconducting phase is well defined, i.e. $\omega_c > \Delta(T,H), T, \Gamma$. Numerous superconductors belong to this extreme type-II family: high-T_c cuprates, A15's, boro-carbides, many organics, etc. The boundaries of the "extreme" pocket, H^* and T^*, defined by $\omega_c \sim \Delta(T=0,H)$ and $\omega_c \sim T$, extend to H as low as $H^* \sim 50\% H_{c2}(0)$ and T as high as $T^* \sim 30\% T_{c0}$ [47]. Within this "extreme" pocket the solution to the superconducting problem must fully incorporate the LL structure of the normal state. This leads to a set of Bogoliubov-de Gennes (BdG) equations for the LL quantized quasiparticles in the presence of the gap function $\Delta(\mathbf{r})$, describing the Abrikosov vortex lattice [3, 4, 5].

The field operators are written as

$$\begin{aligned} \psi_\uparrow(\mathbf{r}) &= \sum_{\nu,\mathbf{q}} \left(u_{\mathbf{q}}^{\nu}(\mathbf{r})\gamma_{\nu,\mathbf{q},\uparrow} - v_{\mathbf{q}}^{*\nu}(\mathbf{r})\gamma_{\nu,\mathbf{q},\downarrow}^{\dagger} \right) \\ \psi_\downarrow(\mathbf{r}) &= \sum_{\nu,\mathbf{q}} \left(u_{\mathbf{q}}^{\nu}(\mathbf{r})\gamma_{\nu,\mathbf{q},\downarrow} + v_{\mathbf{q}}^{*\nu}(\mathbf{r})\gamma_{\nu,\mathbf{q},\uparrow}^{\dagger} \right) \end{aligned} \tag{16}$$

and

$$\epsilon_\beta = \sum_{\sigma,\nu,\mathbf{q}} \epsilon_{\mathbf{q}}^{\nu} \gamma_{\nu,\mathbf{q},\sigma}^{\dagger}\gamma_{\nu,\mathbf{q},\sigma}. \tag{17}$$

Here $\gamma_{\nu,\mathbf{q},\sigma}^{\dagger}$ creates a quasi-particle in the level ν, with momentum $\mathbf{q}$ and spin σ. The problem is diagonal in $\mathbf{q}$ (vectors of the magnetic Brillouin zone) due to the periodicity of the Abrikosov vortex lattice. The amplitudes u and v are expanded in the Landau basis (n is the Landau level) as

$$\begin{aligned} u_{\mathbf{q}}^{\nu}(\mathbf{r}) &= \sum_n u_{\mathbf{q}n}^{\nu}\phi_{\mathbf{q}n}(\mathbf{r}) \\ v_{\mathbf{q}}^{*\nu}(\mathbf{r}) &= \sum_n v_{\mathbf{q}n}^{*\nu}\phi_{-\mathbf{q}n}(\mathbf{r}). \end{aligned} \tag{18}$$

Here $\phi_{\mathbf{q}n}(\mathbf{r})$ are the eigenfunctions of the magnetic translation group in the Landau gauge ($A_x = -Hy, A_y = 0, A_z = 0$) belonging to the nth Landau level. The amplitudes

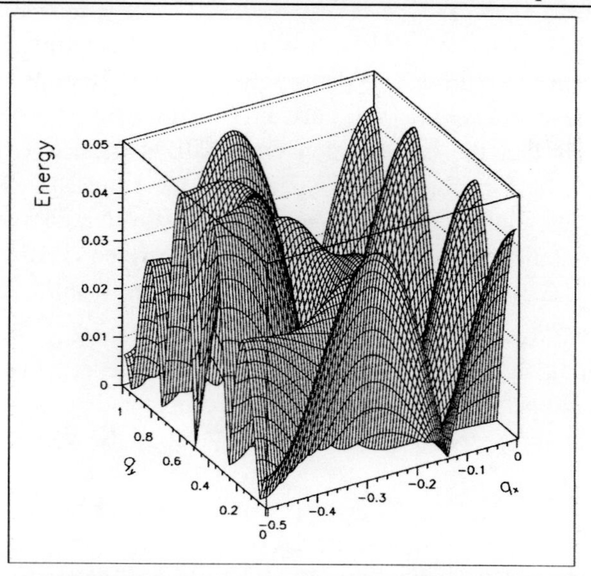

Figure 2: Lowest energy band for $\Delta = 0.1$ in the DA (reproduced from [83]).

$u^\nu_{\mathbf{q}n}$ and $v^\nu_{\mathbf{q}n}$ are the solutions of the BdG equations and $\epsilon^\nu_{\mathbf{q}}$ are the energy eigenvalues. The BdG equations to be solved are

$$\begin{aligned} \epsilon_n u^\nu_{\mathbf{q}n} + \sum_m \Delta_{nm}(\mathbf{q}) v^\nu_{\mathbf{q}m} &= \epsilon^\nu_{\mathbf{q}} u^\nu_{\mathbf{q}n} \\ -\epsilon_n v^\nu_{\mathbf{q}n} + \sum_m \Delta^*_{nm}(\mathbf{q}) u^\nu_{\mathbf{q}m} &= \epsilon^\nu_{\mathbf{q}} v^\nu_{\mathbf{q}n} \end{aligned} \tag{19}$$

where $\epsilon_n = (n+1/2) - \mu/\hbar\omega_c$ (μ is the chemical potential), $\Delta_{nm}(\mathbf{q})$ is the matrix element of the order parameter $\Delta(\mathbf{r})$ between electronic states $(\mathbf{q}, n)$ and $(-\mathbf{q}, m)$ [4].

If the field is high the splitting of the Landau levels is large. Close to the transition to the normal phase the amplitude of the order parameter is small and it is natural that the dominant contribution is to only take the diagonal terms of the matrix Δ_{nm}. In the diagonal approximation (DA) these equations are easilly solved

$$\begin{aligned} \epsilon^\nu_{\mathbf{q}} &= \pm\sqrt{\epsilon_n^2 + |\Delta_{nn}(\mathbf{q})|^2} \equiv \epsilon^n_{\mathbf{q}} \\ u_{\mathbf{q}n} &= \frac{1}{\sqrt{2}}\left(1 + \frac{\epsilon_n}{\epsilon^\nu_{\mathbf{q}}}\right)^{1/2} \\ v_{\mathbf{q}n} &= \frac{1}{\sqrt{2}}\left(1 - \frac{\epsilon_n}{\epsilon^\nu_{\mathbf{q}}}\right)^{1/2} \end{aligned} \tag{20}$$

(and therefore, $\nu \equiv n$). In this case

$$u^\nu_{\mathbf{q}n} v^{*\nu}_{-\mathbf{q}n} = \frac{\Delta_{nn}(\mathbf{q})}{2\epsilon^n_{\mathbf{q}}}. \tag{21}$$

As Δ grows, the off-diagonal terms become increasingly important and the problem has to be solved numerically.

There are two main sources of difficulty in solving these equations. First, the number of LLs involved in the pairing is typically 40-100 and the matrix elements of the gap function between these states, $\Delta_{nm}(\mathbf{q})$, are in general quite complex. A more serious difficulty, however, is that the basis which diagonalizes the BCS Hamiltonian involves *combined* "rotations" *both* in the Nambu space *and* the LL basis. To illustrate why this fact seriously impedes analytic progress note that, to the leading order in $\Delta/\omega_c \ll 1$, the quasiparticle excitation spectrum *near* the Fermi surface (FS) can be found exactly: $E = \pm\sqrt{\varepsilon_n(k_z)^2 + |\Delta_{nn}(\mathbf{q})|^2}$ [4]. It is tempting to generalize this and conclude that the full solution takes the form: $E = \pm\sqrt{\Sigma^2_{\bar{n}\bar{n}} + |D_{\bar{n}\bar{n}}|^2}$, where $\Sigma_{\bar{n}\bar{n}}(\mathbf{q}, k_z)$ and $D_{\bar{n}\bar{n}}(\mathbf{q}, k_z)$ are the normal and pairing self-energies, respectively. However, beyond the leading order, the full solution cannot be put in the above simple form as the "normal" and "pairing" self-energies cannot be *simultaneously* diagonalized. Most of the work beyond leading order has been numerical [4, 48, 5, 49, 50].

In Fig. 2 we show the solution of the energy of the lowest band above the Fermi level, in the DA, as a function of $\mathbf{q}$ for a square lattice (we only show one quadrant of the magnetic Brillouin zone due to the square lattice symmetry) for $\Delta = 0.1$ and $n_c = 10$ (we fix μ at this LL energy). The energy of the lowest band above the Fermi level is given by the gap function $|\Delta_{nn}(\vec{q})|$ in the DA. This is given by [4]

$$\Delta_{nn}(\vec{q}) = \frac{\Delta}{\sqrt{2}} \frac{(-1)^n}{2^{2n} n!} \sum_k e^{2ikq_y a - (q_x + \pi k/a)^2 l^2} H_{2n}[\sqrt{2}(q_x + \pi \frac{k}{a})l] \tag{22}$$

The gap $|\Delta_{nn}(\vec{q})|$ has zeros in the magnetic Brillouin zone in momentum points (q_j) which are in direct correspondence with the positions of the vortices in real space (z_i) such that $q_j l = z_i/l$ [4]. The gapless points grow considerably in number as n_c grows [4]. Including off-diagonal terms small gaps open in the spectrum [51] which are due to the normal (pairing) part of the self-energy and which are of order Δ^2 (Δ^3). However, throughout the Brillouin zone the numerical prefactors are very small and the spectrum is similar almost everywhere to the one obtained in the DA.

These results are best summarized considering the density of states (DOS), defined by

$$\rho(\omega) = \sum_n \left(|u_n|^2 \delta(\omega - E_n) + |v_n|^2 \delta(\omega + E_n)\right) \tag{23}$$

In Fig. 3 we plot the DOS as a function of energy for different values of Δ. For small Δ the DOS is broadened from the Landau level locations by an amount of the order of Δ. The DOS is high at low energies. As Δ grows the energy interval grows and eventually at Δ of the order of the cyclotron energy the lowest band and the next band approach each other and a quantum level-crossing transition occurs. Note that as Δ grows and the DOS spreads over to higher energies, there remains a high DOS at low energies due to the low gap states that retain the characteristic spectrum of the DA. Therefore, it is expected that the DA will be appropriate at low Δ [4, 50] (see however ref. [52] for a thorough discussion). The eigenvalues are grouped in pairs centered around the LL energies. As Δ grows, the bands start to broaden until eventually the gap between them goes to zero and the band-crossing occurs. The sequential eigenvalues have a mirror-like shape with the neighboring bands.

It is however possible to introduce an analytic approach which allows for a transparent and systematic evaluation of corrections to the leading order results [51], for small Δ. The

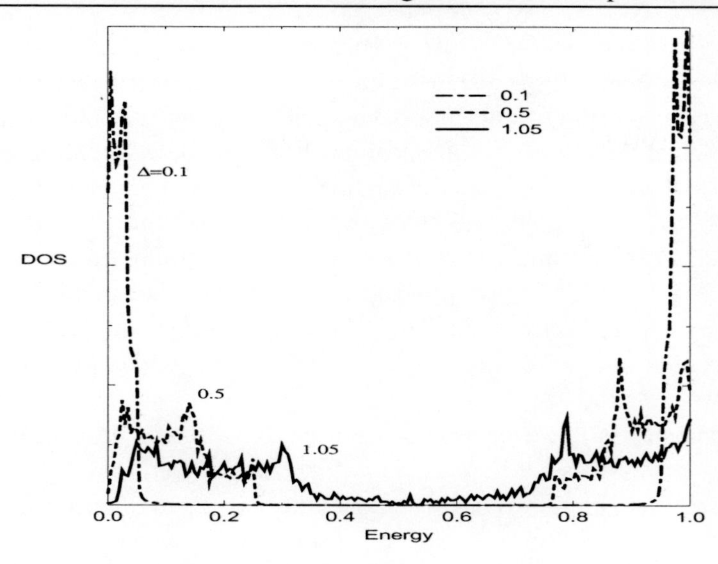

Figure 3: Density of states (DOS) in arbitrary units as a function of energy for $n_c = 10$. The first two bands are shown (reproduced from [83]).

original partition function is replaced by one of "quasiparticles" whose energies have the desired form $E = \pm\sqrt{\tilde{\varepsilon}_{nn}^2 + Z^2|\tilde{\Delta}_{nn}|^2}$, and which faithfully represents the low T (or low energy) properties of the original problem. Here $\tilde{\varepsilon}_{nn}(\mathbf{q}, k_z)$ and $\tilde{\Delta}_{nn}(\mathbf{q}, k_z)$ are the *renormalized* normal and pairing self-energies and Z is the "quasiparticle" renormalization factor.

It was obtained to the leading order in Δ/ω_c that [51]

$$\tilde{\varepsilon}_{nn}(\mathbf{q}, k_z) \approx \varepsilon_n(k_z) + \sum_{p\neq 0} \frac{|\Delta_{n,n+p}(\mathbf{q})|^2}{p\omega_c} \quad . \tag{24}$$

and

$$\tilde{\Delta}_{nn}(\mathbf{q}) = \Delta_{nn}(\mathbf{q}) - $$

$$- \sum_{p\neq 0, p'\neq 0} \frac{\Delta_{n,n+p}(\mathbf{q})\Delta^*_{n+p,n+p'}(-\mathbf{q})\Delta_{n+p',n}(\mathbf{q})}{pp'\omega_c^2} \quad . \tag{25}$$

This "quasiparticle" construction is not an ordinary perturbation theory. It keeps the "normal" and "paring" self-energies separate by postponing the Nambu rotation until the last step. Such representation is valid as long as we are interested in excitations near the Fermi surface.

At the so-called Eilenberger zeroes, $\{\mathbf{q}_i^E\}$ [4] the spectrum *remains gapless*. This statement is correct to *all* orders in perturbation theory and therefore is exact as long as perturbative expansion itself is well-defined. Beyond this point the perturbation theory breaks down due to band crossings between neighboring LL branches. Ultimately, for large Δ/ω_c, there is a crossover to the low-H regime of mini-gapped states in well-separated vortex cores [4], as established by Norman *et al.* [5].

In a s-wave superconductor, isolated vortices contribute a linear term to the low-temperature specific heat. This is due to the core states which have a density of states which can be interpreted as resulting from a normal cylinder of radius the coherence length [6]. In the high field regime the low-T specific heat has been calculated yielding a linear specific heat $C_V = \gamma T$ where $\gamma(H)$ is a nonlinear function of the magnetic field [53]. The results were shown to be in good agreement with experimental results for the superconductor YNi_2B_2C [54]. Also, the comparison of the theory [55] with experimental results for the thermal conductivities yielded a very good agreement for the superconductors $CuNi_2B_2C$ and the $A - 15$ superconductor V_3Si [56]. These results give support to a description of this high-field region in terms of Landau levels, with the consequent gapless and extended nature of the low-lying modes.

The best evidence is however provided by the existence of de Haas-van Alphen oscillations in the mixed phase at high fields [57]. If the lowest states were gapped the de Haas-van Alphen oscillations would be exponentially damped entering the mixed phase, and would not be detected. Since the spectrum is gapless, one can find extended states on the Fermi surface, and the magnetic oscillations persist in the superconducting phase, even though they are attenuated with respect to the normal phase value. Indeed, oscillations were found in $2H - NbSe_2$ [58, 59], in V_3Si [60] and in $YBa_2Cu_3O_{7-\delta}$ [61], and later on in a series of detailed eperiments in V_3Si [47] and Nb_3Sn [62], confirming the existence of the oscillations.

It was found experimentally that the oscillations persist until quite small values of the magnetic field and relatively high values of the temperature. These results were explained theoretically [48, 52] using the Landau level description, which naturally leads to magnetic oscillations. These are attenuated due to the broadening of the Landau levels as one enters the mixed phase, originating in the growth of the gap function. Also, the quasiparticle bands split leading to further attenuation, together with other broadening effects like disorder. The attenuation is however less severe than expected in the vortex lattice case, until the quantum level crossings occur, at which point the oscillations are strongly attenuated [52, 57].

A reasonably good fit of the experimental data for V_3Si was obtained considering a vortex lattice [48] even though other experimental data for other materials are better fitted using a random vortex model [63]. For these systems the attenuation factor is described more accurately using this or equivalent models.

2.5 Quasiparticle-Vortex Scattering

Due to the presence of the vortices the gap function is inhomogeneous (vanishes at the vortex cores) and the quasiparticles scatter off. Also, the singular phase of the gap function leads to circulating currents around the vortices and to Aharonov-Bohm scattering of the quasiparticles off the vortices. Therefore in the presence of vortices the quasiparticles feel the combined effect of the external magnetic field and of the spatially varying field of the chiral supercurrents. The quasiparticles besides feeling a Doppler shift caused by the moving supercurrents [64] (Volovik effect) also feel a quantum "Berry" like term due to a half-flux Aharonov-Bohm scattering of the quasiparticles by the vortices.

In the case of triplet superconductors the transport of the magnetic moment of the pair around the vortex line charge may also lead to an extra phase due to the Aharonov-Casher effect [65]. A similar effect may occur if vortices are allowed to move around electrical charges.

In the case of isolated vortices (or far apart) and considering first s-wave superconductors, the quantum Aharonov-Bohm effect is expected to be small for the lowest energy states. The bound states in the vortex core have a short range and the scattering of the quasiparticles that propagate in the superconductor is not very significant. The quasiparticles are therefore not significantly affected. In this case only the Doppler shift caused by the moving supercurrents will have an important effect on the quasiparticle.

However, in the case of a d-wave vortex, since the low energy states are extended along the nodal directions, an important contribution from regions far from the vortex core is expected, and the Aharonov-Bohm effect will have an important contribution to the quasiparticle scattering. This is expected to be the dominant one. Therefore the d-wave case is particularly interesting.

In the vortex lattice the states are supposedly extended [66, 67]. Therefore the quantum interference effects assume an added importance. In the s-wave case, at least at high fields, where the distance between the vortices is small, we expect important contributions. At lower fields the localized nature of the low energy levels leads back to the single-vortex analysis. In the d-wave vortex lattice the expectation that these are delocalized is evidenced by the increase of the thermal conductivity with field [68] in contrast to the reducing effect of conventional superconductors [69].

The gapless nodes on the Fermi surface in a d-wave superconductor play therefore a determinant role in the low-energy low-temperature properties. It is convenient to linearize the BdG equations around the nodes. It was shown [70] that the eigenfunctions and eigenvectors obey a scaling property related to the magnetic length $l_H \sim H^{-1/2}$. Specifically it was shown that

$$\begin{aligned} \tilde{\psi}_n^H &= \tilde{\psi}_n^{H_0}\left(\vec{r}\left(\frac{H}{H_0}\right)^{1/2}\right) \\ \epsilon_n^H &= \left(\frac{H}{H_0}\right)^{1/2}\epsilon_n^{H_0} \end{aligned} \tag{26}$$

where H and H_0 are two values of the magnetic field. This scaling property, obtained since the spectrum is linear, leads to scaling properties for various physical quantities at low T. Furthermore, neglecting the effect of the superfluid currents, it was proposed that the low energy states have a spectrum of the type [71] $E_n = \pm\sqrt{\omega_c \Delta_0 n}$ where $n = 0, 1, \cdots$, $\omega_c = (eH)/(mc)$ is the cyclotron frequency. These levels result from the quasiparticles precessing in a weak magnetic field along an elliptic orbit resulting from the linearization around a Dirac gap node. However as shown recently [72] the supercurrents strongly mix these Dirac Landau levels and therefore a different spectrum is obtained instead.

In order to solve the problem consider the lattice formulation of a d-wave superconductor in a magnetic field. Let us start from the Bogoliubov-de Gennes equations $\mathcal{H}\psi = \epsilon\psi$ where $\psi^\dagger(\mathbf{r}) = (u^*(\mathbf{r}), v^*(\mathbf{r}))$ and where the matrix Hamiltonian is given by

$$\mathcal{H} = \begin{pmatrix} \hat{h} & \hat{\Delta} \\ \hat{\Delta}^\dagger & -\hat{h}^\dagger \end{pmatrix} \tag{27}$$

with [66, 67]

$$\hat{h} = -t \sum_{\delta} e^{-\frac{ie}{\hbar c} \int_{\mathbf{r}}^{\mathbf{r}+\delta} \mathbf{A}(\mathbf{r}) \cdot dl} \hat{s}_{\delta} - \epsilon_F \tag{28}$$

and

$$\hat{\Delta} = \Delta_0 \sum_{\delta} e^{\frac{i}{2}\phi(\mathbf{r})} \hat{\eta}_{\delta} e^{\frac{i}{2}\phi(\mathbf{r})}. \tag{29}$$

The sums are over nearest neighbors ($\delta = \pm\mathbf{x}, \pm\mathbf{y}$ on the square lattice); $\mathbf{A}(\mathbf{r})$ is the vector potential associated with the uniform external magnetic field $\mathbf{B} = \nabla \times \mathbf{A}$, the operator $\hat{s}_{\delta}$ is defined through its action on space dependent functions, $\hat{s}_{\delta} u(\mathbf{r}) = u(\mathbf{r} + \delta)$, and the operator $\hat{\eta}_{\delta}$ describes the symmetry of the order parameter.

The application of the uniform external magnetic field $\mathbf{B}$ generates in type II superconductors compensating vortices each carying one half of the magnetic quantum flux, $\frac{\phi_0}{2}$. In the London limit, which is valid for low magnetic field and over most of the $H - T$ phase diagram in extreme type-II superconductors like the cuprates, the size of the vortex cores is negligible and each vortex core is placed at the center of a plaquette (unit cell). The N_{ϕ} vortices are distributed over the $L \times L$ plaquettes (randomly if the system is disordered). In this limit one can take the order parameter amplitude Δ_0 constant everywhere in space and factorize the phase of the order parameter.

2.5.1 Franz and Tesanovic Gauge Transformation

At this stage, it is convenient to perform a singular gauge transformation to eliminate the phase of the off-diagonal term (29) in the matrix Hamiltonian. We consider the unitary FT gauge transformation $\mathcal{H} \to \mathcal{U}^{-1} \mathcal{H} U$, where [66]

$$\mathcal{U} = \begin{pmatrix} e^{i\phi_A(\mathbf{r})} & 0 \\ 0 & e^{-i\phi_B(\mathbf{r})} \end{pmatrix} \tag{30}$$

with $\phi_A(\mathbf{r}) + \phi_B(\mathbf{r}) = \phi(\mathbf{r})$. The phase field $\phi(\mathbf{r})$ is decomposed at each site of the two-dimensional lattice in two components $\phi_A(\mathbf{r})$ and $\phi_B(\mathbf{r})$ which are assigned respectively to a set of vortices A, positioned at $\{\mathbf{r}_i^A\}_{i=1,N_A}$, and a set of vortices B, positioned at $\{\mathbf{r}_i^B\}_{i=1,N_B}$. The phase fields $\phi_{\mu=A,B}$ are defined through the equation

$$\nabla \times \nabla \phi_{\mu}(\mathbf{r}) = 2\pi \mathbf{z} \sum_{i} \delta(\mathbf{r} - \mathbf{r}_i^{\mu}) \tag{31}$$

where the sum runs only over the μ-type vortices. After carying out the gauge transformation (30) the Hamiltonian (27) reads

$$\mathcal{H}' = \begin{pmatrix} -t \sum_{\delta} e^{i\mathcal{V}_{\delta}^{A}(\mathbf{r})} \hat{s}_{\delta} - \epsilon_F & \Delta_0 \sum_{\delta} e^{-i\frac{\delta\phi}{2}} \hat{\eta}_{\delta} e^{i\frac{\delta\phi}{2}} \\ \Delta_0 \sum_{\delta} e^{-i\frac{\delta\phi}{2}} \hat{\eta}_{\delta}^{\dagger} e^{i\frac{\delta\phi}{2}} & t \sum_{\delta} e^{-i\mathcal{V}_{\delta}^{B}(\mathbf{r})} \hat{s}_{\delta} + \epsilon_F \end{pmatrix}. \tag{32}$$

The phase factors are given by [67] $\mathcal{V}_{\delta}^{\mu}(\mathbf{r}) = \int_{\mathbf{r}}^{\mathbf{r}+\delta} \mathbf{k}_s^{\mu} \cdot d\mathbf{l}$ and $\delta\phi(\mathbf{r}) = \phi_A(\mathbf{r}) - \phi_B(\mathbf{r})$, where $\hbar \mathbf{k}_s^{\mu} = m\mathbf{v}_s^{\mu} = \hbar \nabla \phi_{\mu} - \frac{e}{c}\mathbf{A}$ is the superfluid momentum vector for the

μ-supercurrent. Note that they are both gauge invariant. Physically, the vortices A are only visible to the particles and the vortices B are only visible to the holes. Each resulting μ-subsystem is then in an effective magnetic field

$$\mathbf{B}_{\text{eff}}^{\mu} = -\frac{mc}{e} \nabla \times \mathbf{v}_s^{\mu} = \mathbf{B} - \phi_0 \mathbf{z} \sum_i \delta^2(\mathbf{r} - \mathbf{r}_i^{\mu}) \tag{33}$$

where each vortex carries now an effective quantum magnetic flux ϕ_0. For the case of a regular vortex lattice [66, 67], these effective magnetic fields vanish simultaneously on average if the magnetic unit cell contains two vortices, one of each type. More generally, in the absence of spatial symmetries, as it is the case for disordered systems, these effective magnetic fields $\mathbf{B}_{\text{eff}}^{\mu=A,B}$ vanish if the numbers of vortices of the two types A and B are equal, i.e. $N_A = N_B$, and their sum equals the number of elementary quantum fluxes of the external magnetic field penetrating the system.

The μ-superfluid wave vector $\mathbf{k}_s^{\mu}(\mathbf{r})$ characterizes the supercurrents induced by the μ-vortices. This vector can be calculated for an arbitrary configuration of vortices [67] like

$$\mathbf{k}_s^{\mu}(\mathbf{r}) = 2\pi \int \frac{d^2k}{(2\pi)^2} \frac{i\mathbf{k} \times \mathbf{z}}{k^2 + \lambda^{-2}} \sum_{i=1}^{\infty} e^{i\mathbf{k}\cdot(\mathbf{r}-\mathbf{r}_i^{\mu})}. \tag{34}$$

The μ-superfluid wave vector distribution is completely determined by the configuration of the μ-vortices and is independent of the pairing symmetry. Since the London limit is taken, where the vortex core size is negligible, any possible different symmetry contributions from inside the vortex core [30] are neglected and the chiral supercurrents simply reflect the circulation of the superfluid density around the vortex. The case $\lambda \to \infty$ is taken, since is an excellent approximation in extreme type-II systems.

The situation where the vortices are regularly distributed in a lattice was treated before [67]. Since the average effective magnetic field vanishes it is possible to solve the BdG equations using a standard Bloch basis– the supercurrent velocities are periodic in space and there is no need to consider the magnetic Brillouin zone. In order to compute the eigenvalues and eigenvectors of the Hamiltonian (32) one looks for eigensolutions in the Bloch form $\Psi_{n\mathbf{k}}^{\dagger}(\mathbf{r}) = e^{-i\mathbf{k}\cdot\mathbf{r}}(U_{n\mathbf{k}}^*, V_{n\mathbf{k}}^*)$ where $\mathbf{k}$ is a point of the Brillouin zone.

Taking the continuum limit and linearizing the spectrum around each node effectively decouples the nodes. It was shown that the low-energy quasiparticles are then naturally described as Bloch waves [66] and not Dirac-Landau levels as previously proposed [71]. The lattice formulation explicitly involves internodal contributions which should be important for the properties of the density of states in the disordered case. In the vortex lattice case, however, it was found that only in special commensurate cases (for the square lattice) the inclusion of the internodal contributions is relevant. In a general incommensurate case the interference leads to qualitatively similar spectra. In the d-wave case the spectrum is gapless with a linear density of states at low energy [67]. One would therefore expect that in a general disordered vortex case internodal scattering might not be relevant (particularly for high Dirac cone anisotropy $\alpha_D = v_F/v_{\Delta} = t/\Delta_0 \gg 1$).

For the conventional s-wave case the operator characterizing the symmetry of the order parameter is constant $\hat{\eta}_{\delta} = \frac{1}{4}$ and the off-diagonal terms of the Hamiltonian (32) are then considerably simplified

$$
\mathcal{H}' = \begin{pmatrix} -t\sum_\delta e^{i\mathcal{V}_\delta^A(\mathbf{r})}\hat{s}_\delta - \epsilon_F & \Delta_0 \\ \Delta_0 & t\sum_\delta e^{-i\mathcal{V}_\delta^B(\mathbf{r})}\hat{s}_\delta + \epsilon_F \end{pmatrix}. \tag{35}
$$

Note that in this case the phase of the off-diagonal term is eliminated.

For the unconventional d-wave case the operator $\hat{\eta}_\delta$ takes the form $\hat{\eta}_\delta = (-1)^{\delta_y}\hat{s}_\delta$, (we recall that $\hat{s}_\delta$ acts on spatial dependent functions as $\hat{s}_\delta u(\mathbf{r}) = u(\mathbf{r}+\delta)$ and that $\delta = \pm\mathbf{x}, \pm\mathbf{y}$ characterizes unit displacements (hops) on the lattice). With these definitions the d-wave Hamiltonian can be derived from the Hamiltonian (32) and reads

$$
\mathcal{H}' = \begin{pmatrix} -t\sum_\delta e^{i\mathcal{V}_\delta^A(\mathbf{r})}\hat{s}_\delta - \epsilon_F & \Delta_0\sum_\delta e^{i\mathcal{A}_\delta(\mathbf{r})+i\pi\delta_y}\hat{s}_\delta \\ \Delta_0\sum_\delta e^{-i\mathcal{A}_\delta(\mathbf{r})-i\pi\delta_y}\hat{s}_\delta & t\sum_\delta e^{-i\mathcal{V}_\delta^B(\mathbf{r})}\hat{s}_\delta + \epsilon_F \end{pmatrix} \tag{36}
$$

where the phase factor $\mathcal{A}_\delta(\mathbf{r})$ has the form

$$
\begin{aligned}
\mathcal{A}_\delta(\mathbf{r}) &= \frac{1}{2}\int_{\mathbf{r}}^{\mathbf{r}+\delta} (\nabla\phi_A - \nabla\phi_B)\cdot d\mathbf{l} \\
&= \frac{1}{2}\int_{\mathbf{r}}^{\mathbf{r}+\delta} \left(\mathbf{k}_s^A - \mathbf{k}_s^B\right)\cdot d\mathbf{l}.
\end{aligned} \tag{37}
$$

In the Hamiltonian (36) and in Eq. 37 the vector

$$
\mathbf{a}_s = \frac{1}{2}\left(\mathbf{k}_s^A - \mathbf{k}_s^B\right) \tag{38}
$$

acts as an internal gauge field independent of the external magnetic field [67]. The associated internal magnetic field $\mathbf{b} = \nabla\times\mathbf{a}_s$ consists of opposite $A-B$ spikes fluxes carying each one half of the magnetic quantum flux ϕ_0, centered in the vortex cores and vanishing on average since the numbers of A- and B-type vortices are the same. In the Doppler shift approximation this term is not taken into account.

The densities of states are shown in Fig. 4 (s-wave case) and in Fig. 5 (d-wave case). In the s-wave case the gap and the boundstates are clearly visible. Also at higher energies the Landau levels are apparent. In the d-wave case there is no gap and the DOS is linear at low ϵ. The influence of the Berry term with respect to the Doppler shift contribution was studied in Ref. [67]. It was clearly shown that the influence of the Berry term significantly changes the structure of the DOS and, therefore approximations where the quantum Berry term is neglected, are not correct.

2.6 DOS, LDOS and STM Measurements

The nature of the states reflects itself in the density of states. This can be measured with a scanning tunneling microscope (STM). In this technique the differential conductance from a metallic tip to the material is proportional to the local density of states (LDOS). This can be controlled varying the distance of the tip to the sample surface or varying the position

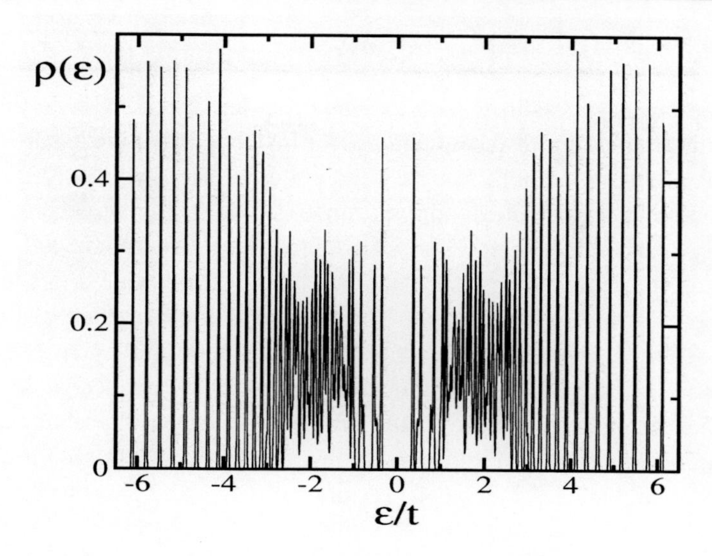

Figure 4: Quasiparticle density of states for the s-wave symmetry for a regular lattice of vortices (reproduced from [21]).

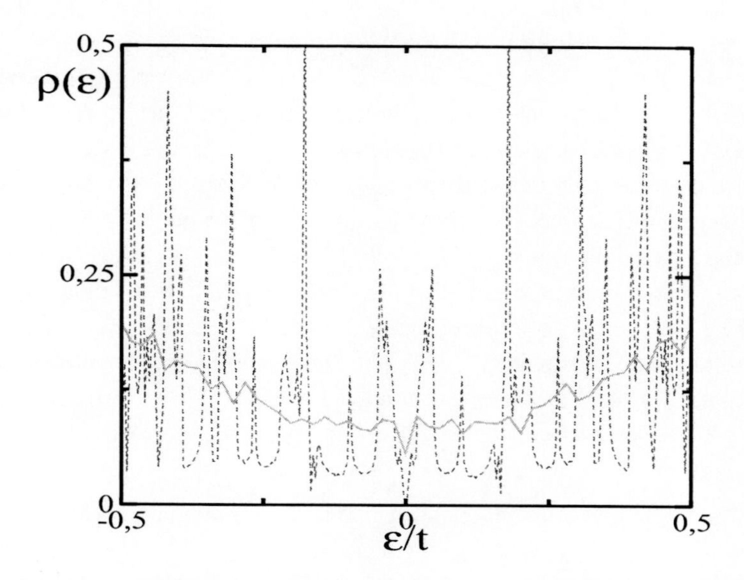

Figure 5: Quasiparticle density of states for the d-wave case without (dashed line) and with disorder (solid line), to be discussed ahead (reproduced from [21]).

of the tip along the plane of the sample. Applying a voltage bias various energies can be probed.

This technique was sucessfully used to detect the density of states in the mixed phase, detecting both the triangular Abrikosov lattice, for instance in a s-wave superconductor like $NbSe_2$ and had precision enough to detect the states inside a vortex [73]. In the case of s-wave superconductors without an applied magnetic field, the density of states shows a gap at zero bias, as expected [73], with two coherence peaks at about the scale of the gap function. Considering small fields, such that the vortices are wide apart, the DOS shows a peak at zero bias at the location of the vortex, due to the presence of the bound states predicted by Caroli, de Gennes and Matricon. Also, as one moves away from the vortex core, the amplitude of the zero bias peak decreases and the coherence peaks start to appear. On a location far from the vortex core the DOS recovers the shape in the absence of a field. Improving the space and energy resolutions it was possible to split the zero bias peak into the discrete bound states [74]. These results were checked and predicted theoretically [27, 34] where it was obtained the detailed structure of the DOS due to the various angular momentum states in the low-T quantum limit [34].

In the case of a d-wave single vortex the gapless nature of the spectrum leads to a zero bias peak in the DOS. However, in this case there are no bound states in the core and the states are extended along the node directions [29].

However, in both cases of s- and d-wave symmetries the presence of the other vortices affects the shape of the LDOS. In the case of a s-wave superconductor in the very high field limit, since its spectrum is gapless, the DOS is qualitatively different with no structure at low energies [4] and no zero bias peak. At intermediate fields the presence of the other vortices leads to a LDOS that has a sixfold star-shaped structure [74]. This result was obtained theoretically in detailed form [75] for s-wave symmetry. At low fields, on the other hand, the vortices are far apart and the LDOS has a cylindrical symmetry around each vortex core (as for a single vortex). As the field increases and the cores start to overlap, the LDOS develops a sixfold star-shaped structure. The ray of the star extends toward the next-nearest-neighbors at low energies and rotates by 30 degrees at high energies such that the ray extends towards the nearest-neighbors [75], as found experimentally. The same effect was found in the case of an atomic crystal field [76].

In the case of d-wave symmetry the presence of the lattice leads to a double peak structure in the LDOS close to a vortex core. For instance in the case of $YBa_2Cu_3O_{7-\delta}$ these peaks are separated by $11meV$ [77]. The same result was also obtained theoretically [38, 67]. Also, the fourfold symmetry predicted for the single vortex is not observed [78]. Later on we will consider the effect of disorder on the density of states and discuss the results in the context of the experimental results.

2.7 Hall Conductance. Thouless Theorem

Several authors have paid attention to the Hall conductance of the mixed phase. In general, there are two contributions: one is due to the vortex motion and the other to the quasiparticle contribution (usually associated with modes localized in the normal region inside the vortex cores). One of the reasons for this interest is that in the superconducting phase σ_{xy} has a different sign with respect to the one of the normal phase [79]. This has been proposed to be due to the vortex motion part since the localized modes are predicted to give a contribution with the same sign as the normal phase value. The case when the vortices crystallize by lowering the temperature was studied [80] yielding interesting behavior for the Hall conductivity of a d-wave superconductor.

In the case of s-wave superconductors in the high-field regime, where Landau quantization is relevant, one expects that the contribution of the quasiparticles to the Hall conductance will be significant. We can consider the vortices as frozen (pinned vortices) and calculate only the contribution due to the quasiparticles.

As mentioned above for large values of Δ a regime is reached where the quasiparticles are bound to the vortex cores with an energy spacing typical of an isolated single vortex [52]. In this regime one expects that a tight-binding approximation should yield good results, since in this lower field regime the vortex cores are sparse. Note that as Δ increases the magnetic field decreases. The relevant scale is the ratio Δ/ω_c. Such a procedure has been carried out and a gapped regime is indeed found [52, 81] signalling a transition from the high-field gapless regime to the low-field gapped regime. Between the small Δ regime and this gapped regime several quantum level-crossings take place.

On the other hand, Thouless has shown that the Landau levels do not constitute a complete basis to be used in a tight-binding approximation scheme and therefore it is not possible to construct well localized representations of the magnetic translation group unless the Hall conductance, σ_{xy}, is zero [82]. Since it is intuitive that, in the limit the vortex concentration is low, a tight-binding description should be appropriate, we expect, in the light of Thouless theorem, that σ_{xy} should be zero in this low-field regime. This is suggested by the localized nature evidenced by the numerical solution of the BdG equations. It was argued [83] that the Hall conductance could be used as an order parameter to signal the transition from the high-field gapless regime, where Landau quantization has been shown to occur, (finite σ_{xy}) to the low-field gapped regime (zero σ_{xy}).

The quasiparticle contribution to the Hall conductance [83], considering a pinned vortex lattice, can be calculated using the Kubo formula

$$\begin{aligned}\sigma_{xy}(\mathbf{r},\mathbf{r}') &= -i\hbar L^2 \sum_{\beta\neq 0}\{<0|J_x(\mathbf{r})|\beta><\beta|J_y(\mathbf{r}')|0> \\ &- <0|J_y(\mathbf{r}')|\beta><\beta|J_x(\mathbf{r})|0>\}\frac{1}{(\epsilon_\beta-\epsilon_0)^2}\end{aligned} \tag{39}$$

where the currents are given by

$$\begin{aligned}J_i(\mathbf{r}) &= \frac{e\hbar}{2imc}\sum_\sigma\{\psi_\sigma^\dagger(\mathbf{r})\left(\frac{\partial}{\partial x_i}\psi_\sigma(\mathbf{r})\right) \\ &- \left(\frac{\partial}{\partial x_i}\psi_\sigma^\dagger(\mathbf{r})\right)\psi_\sigma(\mathbf{r})\} - \frac{e^2}{mc^2}A_i\sum_\sigma\psi_\sigma^\dagger(\mathbf{r})\psi_\sigma(\mathbf{r})\end{aligned} \tag{40}$$

Here σ is the spin projection, $i = x, y, z$, $\mathbf{A}$ is the vector potential, $\psi_\sigma(\mathbf{r})$ is the electron field operator, the energies ϵ_β are the full solution of the many-body problem and $\beta = 0$ is the groundstate. We consider a square lattice of side L and calculate the Hall conductance obtaining the energies from the solution of the BdG equations.

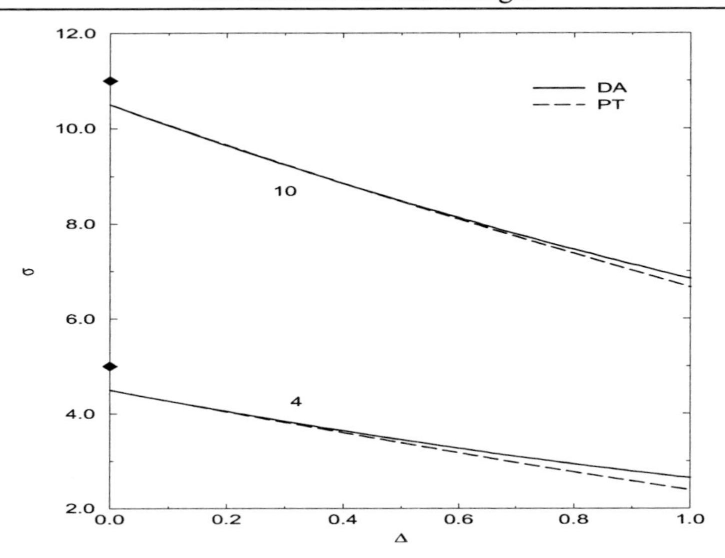

Figure 6: Hall conductance for $n_c = 4, 10$ as a function of Δ for the diagonal approximation (DA) and leading order perturbation theory (PT). The two points at $n_c + 1$ are the $\Delta = 0$ result (normal phase) (reproduced from [83]).

In the DA the expression for the Hall conductance σ takes the form

$$\begin{aligned} \frac{\sigma}{\frac{q^2}{h}} &= -\frac{1}{4}\sum_{\beta\neq 0}\frac{1}{(\epsilon_\beta-\epsilon_0)^2}\frac{1}{N_\phi}\sum_{\mathbf{q}}\sum_{n} \\ & \Re\{(n_{n\mathbf{q}}n_{n+1,\mathbf{q}})'(n+1)[\frac{\Delta^*_{nn}}{\epsilon^n_{\mathbf{q}}}\frac{\Delta_{n+1,n+1}}{\epsilon^{n+1}_{\mathbf{q}}} - \frac{\epsilon^n_{\mathbf{q}}-\epsilon_n}{\epsilon^n_{\mathbf{q}}}\frac{\epsilon^{n+1}_{\mathbf{q}}+\epsilon_{n+1}}{\epsilon^{n+1}_{\mathbf{q}}}] \\ & -(n_{n\mathbf{q}}n_{n-1,\mathbf{q}})'(n)[\frac{\Delta^*_{nn}}{\epsilon^n_{\mathbf{q}}}\frac{\Delta_{n-1,n-1}}{\epsilon^{n-1}_{\mathbf{q}}} - \frac{\epsilon^n_{\mathbf{q}}-\epsilon_n}{\epsilon^n_{\mathbf{q}}}\frac{\epsilon^{n-1}_{\mathbf{q}}+\epsilon_{n-1}}{\epsilon^{n-1}_{\mathbf{q}}}]\} \end{aligned} \tag{41}$$

Here $\epsilon_\beta = \epsilon^n_{\mathbf{q}} + \epsilon^{n+1}_{\mathbf{q}}$ in the first term and $\epsilon_\beta = \epsilon^n_{\mathbf{q}} + \epsilon^{n-1}_{\mathbf{q}}$ in the second term. Fixing the chemical potential at the level n_c ($n_c + 1$ occupied levels) and taking $\Delta = 0$ (normal phase) we get that $\bar{\sigma} = \sigma h/q^2 = n_c + 1$, as expected. The only contribution comes from the first term with $n = n_c$. Taking now a small value of Δ the two dominant contributions come from the previous term and from the first term with $n = n_c - 1$. This leads to a discontinuity in σ: at $\Delta = 0$ the Hall conductance $\bar{\sigma} = (n_c + 1)$ while at small, but nonzero Δ, $\bar{\sigma} \sim (n_c + 1) - 1/2$. As Δ grows $\bar{\sigma}$ decreases continuously.

In Fig. 6 we show $\bar{\sigma}$ as a function of Δ for the DA and for the leading order perturbation theory (PT) eqs. (27) and (28) in the range of values of Δ up to the order of the first level-crossing. We consider the cases $n_c = 4, 10$. As Δ grows the off-diagonal terms renormalize downwards the Hall conductance with respect to the DA value. In the normal phase ($\Delta = 0$) the Fermi level (for a completely filled level) is in the gap between two Landau levels. As Δ is turned on, μ is kept fixed at $\mu = n_c + 1/2 + \eta$ ($\eta \to 0$) and the levels spread to higher energies. This leads to a decrease of the Hall conductance since the fraction of low energy states that may conduct decreases. The presence of the off-diagonal terms fastens the rate of decrease for larger values of Δ, for the same reason. At very low Δ the two methods agree, as expected. Close to the level crossing the study is very difficult because a fully self-consistent calculation is needed to properly cover the crossover to the gapped regime. Going beyond the DA the several eigenfunctions of the BdG equations have components in the various Landau indices, which due to the gapless (or almost gapless) nature of the spectrum causes difficult numerical problems due to the energy denominator in eq. (50). Also, strong mixing of the LL destroys the LL structure beyond the level-crossing(s) transition(s). There is a discontinuity in the Fermi level due to this level crossing (by $\pm 2\omega_c$ because of the doubling due to the particle and hole bands, u and v, respectively). After the transition, $\bar{\sigma}$ is expected to decrease again and to tend to zero as Δ grows even further (eventually after several level-crossings). The numerical procedure becomes unreliable for large values of Δ and a different procedure would have to be adopted.

We can as well calculate the Hall conductance directly at intermediate fields. Consider the lattice formulation previously introduced. At zero temperature the Hall conductance can be calculated from

$$\sigma_{\alpha\beta} = \frac{1}{\omega V} \int_0^\infty dt < 0|[j_\alpha^\dagger(0,t), j_\beta(0,0)]|0 > e^{i\omega t} \tag{42}$$

where $j(t) = e^{iHt} j e^{-iHt}$ is the current operator and $|0>$ is the groundstate. Inserting a decomposition of the identity and considering the zero frequency limit it reduces to

$$\sigma_{\alpha\beta} = -\frac{i}{V} \sum_M \frac{< 0|j_\alpha^\dagger|M >< M|j_\beta|0 > - < 0|j_\beta|M >< M|j_\alpha^\dagger|0 >}{\epsilon_M^2 + \eta^2} \tag{43}$$

where $\epsilon_M = E_M - E_0$ is the energy difference between the excited state $|M>$ and the groundstate. In the presence of a field the lattice current operator can be obtained from $\vec{j} = \sum_{\vec{r}} \vec{j}(\vec{r})$ with

$$\vec{j}(\vec{r}) = -it \sum_{\vec{\delta},\sigma} \vec{\delta} \psi^\dagger_{\vec{r}+\vec{\delta},\sigma} \psi_{\vec{r},\sigma} e^{-\frac{ie}{\hbar c} \int_{\vec{r}}^{\vec{r}+\vec{\delta}} \vec{A}(\vec{r})\cdot d\vec{l}} \tag{44}$$

where $\vec{\delta}$ is the vector connecting to the neighbors of the point $\vec{r}$ and t is the hopping. Using the Franz, Tesanovic gauge transformation the field operators are replaced by

$$\begin{pmatrix} \alpha_\uparrow(\vec{r}) \\ \alpha_\downarrow^\dagger(\vec{r}) \end{pmatrix} = \begin{pmatrix} e^{-i\Phi_A} \psi_\uparrow(\vec{r}) \\ e^{i\Phi_B} \psi_\downarrow^\dagger(\vec{r}) \end{pmatrix}$$

and the hermitian conjugates. The current operator is then written as

$$\vec{j} = -it \sum_{\vec{r},\vec{\delta}} \vec{\delta} \{ e^{-i\mathcal{V}_\delta^A(\vec{r})} \alpha_\uparrow^\dagger(\vec{r}+\vec{\delta}) \alpha_\uparrow(\vec{r}) + e^{-i\mathcal{V}_\delta^B(\vec{r})} \alpha_\downarrow^\dagger(\vec{r}+\vec{\delta}) \alpha_\downarrow(\vec{r}) \} \tag{45}$$

Performing the Bogoliubov-Valatin transformation

$$\begin{aligned} \alpha_\uparrow(\vec{r}) &= \sum_n \left(\tilde{u}_n(\vec{r}) \gamma_{n\uparrow} - \tilde{v}_n^*(\vec{r}) \gamma_{n\downarrow}^\dagger \right) \\ \alpha_\downarrow(\vec{r}) &= \sum_n \left(\tilde{u}_n(\vec{r}) \gamma_{n\downarrow} + \tilde{v}_n^*(\vec{r}) \gamma_{n\uparrow}^\dagger \right) \end{aligned} \tag{46}$$

the Hall conductance can be obtained from

$$\sigma_{\alpha\beta} = -i\frac{1}{V}\sum_{M\neq 0}\frac{1}{(\epsilon_n+\epsilon_m)^2+\eta^2}\sum_{\vec{r},\vec{\delta}}\sum_{\vec{r}',\vec{\delta}'}\sum_{n,m}\delta_\alpha\delta'_\beta$$
$$\{\ T^1_{nm}(\vec{r})^* T^1_{nm}(\vec{r}') - T^2_{nm}(\vec{r})^* T^2_{nm}(\vec{r}')\} \qquad (47)$$

where $E_M - E_0 = \epsilon_n + \epsilon_m$ and

$$T^1_{nm}(\vec{r}) = -te^{-i\mathcal{V}^A_{\vec{\delta}}(\vec{r})}u^*_n(\vec{r}+\vec{\delta})v^*_m(\vec{r}) - te^{-i\mathcal{V}^B_{\vec{\delta}}(\vec{r})}u^*_m(\vec{r}+\vec{\delta})v^*_n(\vec{r})$$
$$T^2_{nm}(\vec{r}) = te^{-i\mathcal{V}^B_{\vec{\delta}}(\vec{r})}v_n(\vec{r}+\vec{\delta})u_m(\vec{r}) + te^{-i\mathcal{V}^A_{\vec{\delta}}(\vec{r})}v_m(\vec{r}+\vec{\delta})u_n(\vec{r}) \qquad (48)$$

and therefore $\sigma_{\alpha,\beta} = 0$. This result is consistent with the expectation that in this regime the Hall conductance should vanish. However, it remains to be determined the point where the Hall conductance goes to zero as the magnetic field is lowered, in particular if it coincides with the change to the regime where the CGM states appear and to determine if the transition is second order or first order.

2.8 Spin Hall Conductance

In the normal phase and in the presence of a strong magnetic field perpendicular to a quasi-two dimensional system, the (charge) Hall conductance is quantized (in the presence of disorder). Since the charge is ill-defined in a superconductor due to the mixture of particle and hole states (or since the Cooper pairs have charge $2e$) the Hall conductance in the superconducting phase is not quantized. However, both the spin and the energy are quantized and it is expected that the spin Hall conductance and the thermal Hall conductivity are quantized [84, 85] at zero temperature. The transverse spin Hall conductance is quantized in units of $\hbar/8\pi$ if the spectrum is gapped. This result follows from a topological consideration in a way similar to the quantization of the charge Hall conductance in the quantum Hall effect. The Wiedemann-Franz law relates the thermal Hall conductivity to the spin Hall conductance yielding also a quantized value for K_{xy}/T.

The gapped spectrum occurs in various cases like an s-wave superconductor in not very high field, in a d-wave superconductor with $d_{x^2-y^2}+id_{xy}$ symmetry or in a regular d-wave superconductor in a inversion-symmetric vortex lattice.

3 Effects of Disorder and Magnetic Field

3.1 High Field Landau Regime: Impurity Disorder

In the high field regime the case of a dirty but homogeneous s-wave superconductor was considered before [17]. It was assumed that the order parameter is not significantly affected by the impurities and retains its periodic structure. It was found that at low energies the density of states behaves as $\rho(\epsilon) \sim \epsilon^2$ for small values of the impurity concentration and scattering length (weak scattering potentials). As the disorder becomes stronger, by increasing the impurity concentration, the density of states becomes finite at zero energy. This implies extended gapless zones due to the presence of disorder. As we will see later this is a common effect of the random disorder at low energies.

3.2 High Field Landau Regime: Vortex Disorder

In general, since the interactions between the vortices are repulsive, a lattice structure is more favorable energetically. However, if pinning centers are present in the system, the higher energies of different configurations of the vortices may be offset by the presence of disorder. In this case of strong pinning it is interesting to calculate the effect of disorder on the density of states. Also, a question that arises is if the gapless behavior found in the lattice case at high fields prevails in this more general case. We neglect at this point the scattering off impurities studied above, separating therefore the two contributions.

If the magnetic field is very high, such that the system is in the quantum limit where the electrons are confined to the lowest Landau level, it has been shown that for an arbitrary configuration of zeros there is at least one gapless point [86]. A similar result holds for not so high fields.

Let us consider once again the real space BdG equations If $\Delta(\vec{r}) = 0$, the solutions are the Landau eigenfunctions which, for a two-dimensional system perpendicular to the magnetic field, read in the Landau gauge

$$\phi_{nq} = \frac{1}{\sqrt{L_x}} \frac{1}{\sqrt{l\sqrt{\pi}2^n n!}} e^{iqx} e^{-\frac{1}{2}\left[\frac{y}{l}+ql\right]^2} H_n \left[\frac{y}{l} + ql\right] \tag{49}$$

where n is the Landau index, L_x is the length of the system in the x direction, l is the magnetic length and H_n is an Hermite polynomial. The energy eigenvalues are those of an harmonic oscillator centered at $y_0 = -ql^2$, $E_n = \hbar\omega_c \left(n + \frac{1}{2}\right)$, where ω_c is the cyclotron frequency. Taking L_y to be the dimension along y we obtain that $-L_y/2 \leq ql^2 \leq L_y/2$. In the presence of $\Delta(\vec{r}) \neq 0$ we can use the Landau basis like

$$\begin{aligned} u(\vec{r}) &= \sum_{nq} u_{nq}\phi_{nq}(\vec{r}) \\ v(\vec{r}) &= \sum_{nq} v_{nq}\phi^*_{nq}(\vec{r}) \end{aligned} \tag{50}$$

and obtain the corresponding eigensystem

$$\begin{aligned} \left[n - n_c\right] u^\mu_{nk} + \sum_{mq} v^\mu_{mq}\Delta^{kq}_{nm} &= \epsilon^\mu u^\mu_{nk} \\ -\left[n - n_c\right] v^\mu_{nk} + \sum_{mq} u^\mu_{mq}\left(\Delta^{qk}_{mn}\right)^* &= \epsilon^\mu v^\mu_{nk} \end{aligned} \tag{51}$$

where $\epsilon^\mu = E^\mu/(\hbar\omega_c)$, n_c is defined by $\mu = \hbar\omega_c(n_c + \frac{1}{2})$ and

$$\Delta^{kq}_{nm} = \int d\vec{r}\phi^*_{nk}(\vec{r})\frac{\Delta(\vec{r})}{\hbar\omega_c}\phi^*_{mq}(\vec{r}). \tag{52}$$

The excitation spectrum is then obtained with an appropriate choice for the order parameter.

In the lattice case, Abrikosov's solution can be written in the Landau gauge as [1]

$$\Delta(\vec{r}) = \Delta \sum_p e^{i\pi\frac{b_x}{a}p^2} e^{i\frac{2\pi p}{a}x} e^{-\left(\frac{y}{l}+\frac{\pi p}{a}l\right)^2} \tag{53}$$

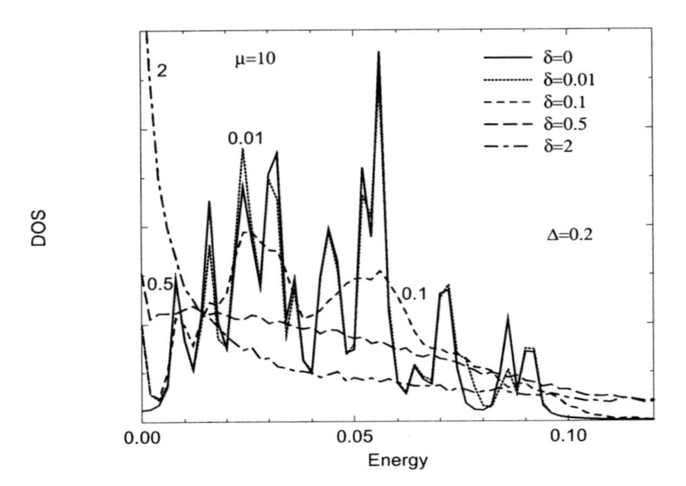

Figure 7: Density of states (DOS) (in arbitray units) as a function of energy for $n_c = 10 \pm 2$, for $\Delta = 0.2$ for the cases of $\delta = 0, 0.01, 0.1, 0.5, 1, 2$. Only the lowest band is shown (reproduced from [18]).

The vortex lattice is characterized by unit vectors $\vec{a} = (a, 0)$ and $\vec{b} = (b_x, b_y)$ where $b_x = 0$, $b_y = a$ for a square lattice and $b_x = \frac{1}{2}a$, $b_y = \frac{\sqrt{3}}{2}a$ for a triangular lattice. This form for the order parameter is valid sufficiently close to the upper critical field, since it is entirely contained in the lowest Landau level of Cooper charge $2e$. In this case the self-consistent equation for the order parameter reduces to a single relation between Δ and V, the attractive interaction strength between the electrons. In the following we will consider the square lattice ($L_x = L_y = L$), for simplicity. In this case the lattice constant $a = l\sqrt{\pi}$ and the zeros are located at the points $x_i = (i + \frac{1}{2})l\sqrt{\pi}$, $y_j = (j + \frac{1}{2})l\sqrt{\pi}$.

An expression for the order parameter has also been found for an arbitrary distribution of the zeros which, in the symmetric gauge, can be written as [87]

$$\Delta(x, y) = \bar{\Delta} \prod_{i=1}^{N_\phi} \left[\frac{x - x_i}{l} + i \frac{y - y_i}{l} \right] e^{-\frac{1}{2l^2 N_\phi}\left[(x-x_i)^2 + (y-y_i)^2\right]} \tag{54}$$

Here, N_ϕ is the number of vortices in the system (number of zeros (x_i, y_i)). The solution of the BdG equations is then obtained numerically performing the gauge transformation to the Landau gauge.

In the lattice case it is more convenient to use a representation in terms of the magnetic wave-functions [88, 4], to take advantage of the translational invariance. The generalization to a random configuration is however more conveniently done using a real space representation. It is convenient to define new variables $X = \frac{x}{L}$, $Y = \frac{y}{L}$ and to use that $L = l\sqrt{\pi N_\phi}$. Considering a finite size system and using periodic boundary conditions calculating $\Delta(x, y)$ over a set of zeros contained in a circle of unit radius (recall that $-\frac{1}{2} \leq X, Y \leq \frac{1}{2}$) the excitation spectrum and the eigenvectors can be obtained and the density of states can be calculated.

Three cases were considered [18]: i) square lattice, ii) weakly disordered lattice and iii) randomly pinned configuration (strong disorder). In the case of weak disorder the distribution of zeros is defined by $X_i = X_i^L + \delta(-\frac{1}{2} + r)/\sqrt{N_\phi}$, where X_i^L is the regular lattice location, δ is an adjustable amplitude and r is a random number $0 < r < 1$ (and similarly for Y_i). In the strong disorder case X_i, Y_i can take any values on the system. The average over disorder is calculated directly on the density of states. One randomly selects one configuration of the zeros and obtains the excitation spectrum and $\rho(\omega)$. The process is repeated many times and the final result is obtained taking the average over the resulting expressions for the density of states.

In Fig. 7 we consider the case of $n_c = 10$ and $\mu = 10$ as a function of disorder. The case $\delta = 0$ shows a small finite density of states at the Fermi level (due to the finite size considered) and a large density of states at low energies. As the disorder increases its effect is very pronounced. The DOS broadens and extends from zero energy with a zero energy value that increases as the δ increses. For strong disorder ($\delta = 2$) the DOS is much larger than the lattice result. The case of $\delta = 2$ is already very similar to the case of strong disorder where the randomness is maximized. Even though the disorder strongly affects the $u(\vec{r})$ and $v(\vec{r})$ amplitudes, no evidence was found for localization. The lowest energy eigenvectors extend considerably over the whole system even though they are strongly inhomogeneous.

The results show that in the presence of disorder the gapless behavior does not disappear and is actually "enhanced". The numerical results indicate that there is a finite density of states at zero energy. They also confirm that the gapless behavior has a topological nature and is not specific to the periodic vortex lattice.

3.3 Random Vortices in Low to Intermediate Fields

Consider now the BdG equations in low to intermediate fields. We consider the disorder induced by the random pinning of the N_ϕ vortices over the two-dimensional $L \times L$ lattice. As the effective magnetic fields experienced by the particles and the holes vanish on average within the Franz-Tesanovic gauge transformation, one is allowed to use periodic boundary conditions on the square lattice ($\Psi(x+nL, y) = \Psi(x, y+mL) = \Psi(x, y)$ with $n, m \in Z$). The $L \times L$ original lattice becomes then a magnetic supercell where the vortices are placed at random with a mean intervortex spacing $\ell = 1/\sqrt{B}$. The disorder in the system is then established over a length L. The Hamiltonian $e^{-i\mathbf{k}\cdot\mathbf{r}}\mathcal{H}' e^{i\mathbf{k}\cdot\mathbf{r}}$ is diagonalized for a large number of points $\mathbf{k}$ in the Brillouin zone and for many different realizations (around 100) of the random vortex pinning.

Consider first the case of a disordered s-wave superconductor. The gap is filled by the disorder. In contrast, the Landau level quantization structure clearly persists at high excitation energies. In general the increase of the magnetic field modifies the curvature of the quasiparticle density of states for the low energies. As it is shown in Fig. 8 the density of states seems to be decribed by the power-law formula

$$\rho(\epsilon) \sim \epsilon^\alpha. \tag{55}$$

This exponent obeys the law $\alpha \simeq c\ell - d$, where $\ell = 1/\sqrt{B}$ is the mean intervortex spacing and d is found close to 1. In the cases of a strong magnetic field the density of vortices is high and strictly we are in a regime where the size of the vortex cores can not be

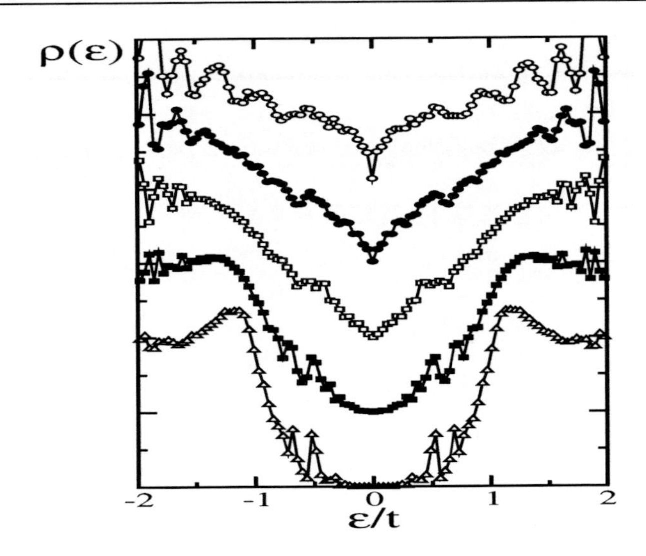

Figure 8: Quasiparticle density of states (s-wave) for different magnetic fields $B = 2/200$ (triangle), $B = 4/200$ (filled square), $B = 7/200$ (square), $B = 11/200$ (bullet) and $B = 20/200$ (circle) in units of $hc/(2e\delta^2)$. The linear system size is $L = 20$ and the parameters are $\mu = -2.2t$, $\Delta_0 = t$. For clarity the different curves are vertically shifted (reproduced from [21]).

neglected. In this regime the Landau level structure at high energies is clear and it extends to low energies superimposed by the effects of disorder.

Consider now the d-wave case. At weak fields the density of states is small at low energies having a dip close to zero energy. Consider the isotropic case $\alpha_D = 1$. The density of states at small energies is finite up to quite small energies where there is a dip to a value that decreases as the magnetic field decreases. Only for quite small magnetic fields the density of states approaches zero at the origin. Neglecting the narrow region close to the origin the density of states can be fitted using the power law

$$\rho(\epsilon) = \rho_0 + \beta|\epsilon|^\alpha. \tag{56}$$

Reasonable fits were obtained taking $\alpha \sim 2$ and we obtain that $\rho_0 \sim B^{1/2}$. The various system sizes fit in the same universal curve indicating that the finite size effects are negligible. In the lattice case the density of states at low energies is linear [66, 89, 67, 90] (this result differs from the behavior obtained by other authors for a d-wave superconductor with no disorder [64, 91], where $\rho(\epsilon \sim 0) \sim B^{1/2}$). The finite density of states at zero energy is therefore a consequence of finite disorder.

Focusing on the narrow region close to $\epsilon = 0$, except for the very low fields, such that only two vortices pierce the disordered supercell, the density of states seems to be finite at zero energy. As shown in Ref. [89] in this case the spectrum is usually gapped and therefore the density of states vanishes at zero energy. Performing a fit like in Eq. 56 one obtains an exponent which is now close to 1. In this regime $\rho_{0\text{dip}}$ also scales with $\sqrt{B}$ and the slope scales linearly with B. At these low energies the density of states for the various system

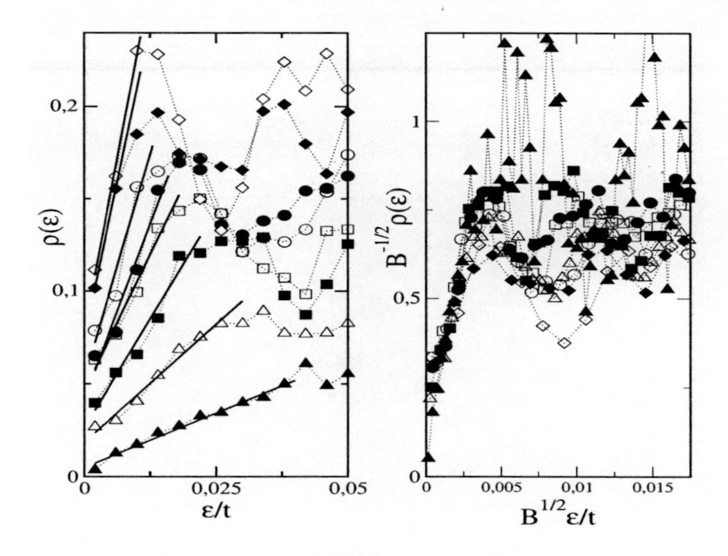

Figure 9: Low-energy quasiparticle density of states $\rho(\epsilon)$ for the d-wave case and for different magnetic fields from $B = 1/200$ (filled triangle), to $B = 25/200$ (lozenge). In the right panel we present the near scaling at low energies (reproduced from [21]).

sizes appears to be of the following approximate form

$$\rho(\epsilon) \sim \frac{1}{\omega_H}\frac{1}{l^2}\mathcal{F}\left(\frac{\epsilon}{\omega_H}\frac{\delta^2}{l^2}\right), \tag{57}$$

where $\omega_H \sim \sqrt{\Delta B}$ and $\mathcal{F}$ is a universal function. In the left panel of Fig. 9 $\rho(\epsilon)$ is shown for various fields in units of $hc/(2e\delta^2)$, while in the right panel it is illustrated the near scaling at low energies consistent with $\mathcal{F}(x) \sim c_1 + c_2 x$ at small x. The density of states is fitted like $\rho(\epsilon) = \rho_{0,dip} + \beta|\epsilon|$. Note that $\beta(B \to 0)$ appears to be small but finite, consistent with a crossover to a "Dirac node" scaling $\rho(\epsilon) \sim (1/\omega_H)(1/l^2)\mathcal{F}(\epsilon/\omega_H)$ at very low fields.

It is interesting to study the dependence of the quasiparticle density of states on the value of the Dirac anisotropy ratio α_D. Such a study is interesting in order to compare the results with experiments; indeed for high-T_c superconductors such as $YBa_2Cu_3O_7$ and $Bi_2Sr_2CaCu_2O_8$ the Dirac anisotropy ratio is [68, 92] respectively $\alpha_D \simeq 14$ and $\alpha_D \simeq 19$.

In the lattice case a high anisotropy increases the density of states at low energies and leads to lines of quasi-nodes [67, 89]. At high anisotropy ($\Delta_0 << t$) the nodes are very narrow and a one-dimensional like character is evidenced. The low-energy quasiparticle density of states at constant field is filled when the anisotropy parameter α_D is increased. There is a narrow linear region close to the origin that decreases as α_D increases. It turns out that the low-energy quasiparticle density of states has the form

$$\rho(\epsilon) \sim \rho_0 + \Gamma\left(\frac{\epsilon}{\Delta_0}\right)^{\alpha} \tag{58}$$

with zero-energy density of states $\rho_0 \sim \sqrt{\frac{t}{\Delta_0}}$, the exponent $\alpha \simeq 1+\sqrt{\frac{\Delta_0}{t}}$ and the parameter $\Gamma \sim \frac{\Delta_0}{t}$.

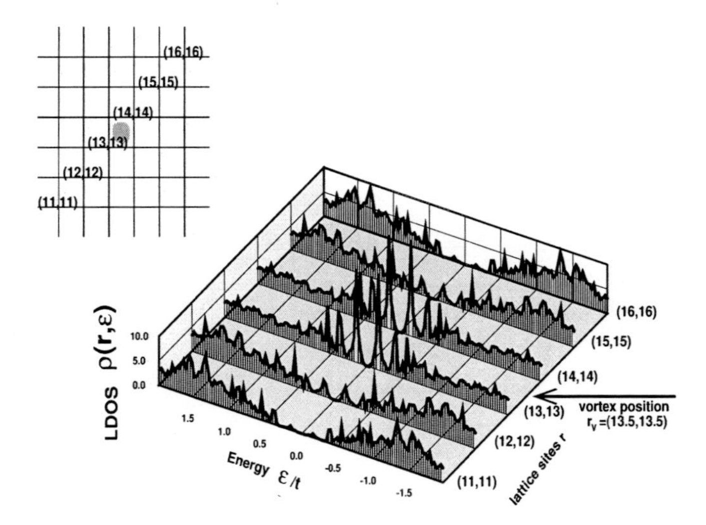

Figure 10: Quasiparticle local density of states $\rho(\mathbf{r}, \epsilon)$ for the d-wave case at different lattice sites around a vortex position belonging to a regular vortex lattice (reproduced from [21]).

The dependence of the exponent is interesting since in the isotropic case we retrieve $\alpha = 2$ and with increasing anisotropy the exponent decreases to 1 characteristic of the Dirac nodes. At $\epsilon \simeq 0$ and for a high Dirac anisotropy ratio the quasiparticle density of states flattens since the coefficient Γ in (58) decreases with Δ_0.

An analysis of the nature of the states showed [21] that the states are extended, even though the system is disordered, particularly if the energies are not very small. In this regime the results were also consistent with extended states even though finite size effects at very low energies could not be ruled out entirely [21].

At low energies the LDOS defined by

$$\rho(\mathbf{r}, \epsilon) = \sum_{\mathbf{n}} \Big(|u_{\mathbf{n}}(\mathbf{r})|^2 \, \delta(\epsilon - E_{\mathbf{n}}) + |v_{\mathbf{n}}(\mathbf{r})|^2 \, \delta(\epsilon + E_{\mathbf{n}}) \Big) \tag{59}$$

is strongly peaked near the vortex cores. At higher energies ($\epsilon \simeq 1$) the LDOS is much more homogeneously spread over the system indicative of extended states; the values of the LDOS over the lattice are approximatively constant and slightly fluctuate around the expected value for extended states.

A spatial scanning of the LDOS in the vicinity of a vortex core for the case of a regular vortex lattice is shown in (Fig. 10) and for the case of a disordered vortex lattice in (Fig. 11). For both cases the LDOS away from the vortex core are qualitatively the same and are comparable to that of a d-wave superconductor in a zero-magnetic field. Also, for both cases the low-lying states are predominant in the close vicinity of the vortex core but their respective spectra are different. For the regular vortex lattice case (Fig. 10) there is a double peak structure around the vortex core. This result is in qualitative agreement with the double peak in conductance observed in $YBa_2Cu_3O_{7-\delta}$ by scanning tunneling microscopy (STM) [77]. For the disordered vortex case (Fig. 11) zero energy peaks (ZEP) appear in the close vicinity of the vortex core (closest neighbor sites) and rapidily disappear (typically over 3δ)

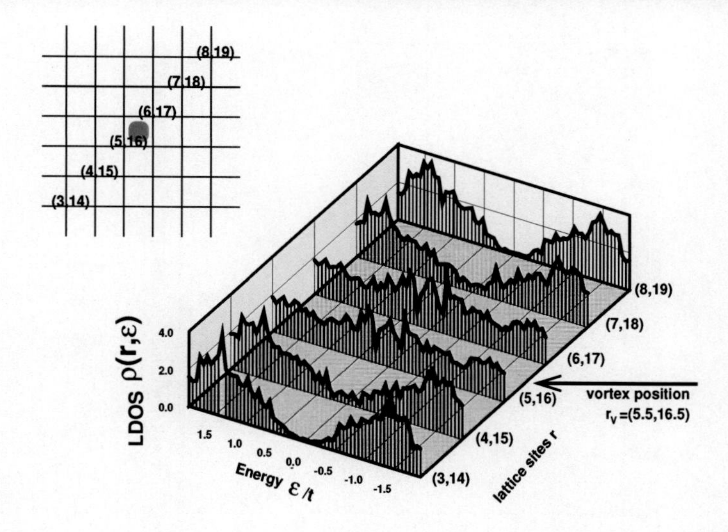

Figure 11: Quasiparticle local density of states $\rho(\mathbf{r}, \epsilon)$ for the d-wave case at different lattice sites around a vortex position belonging to a disordered vortex lattice (reproduced from [21]).

when moving away from the vortex core. We note also that close to the vortex cores the coherence peaks disappear in both the regular and the disordered cases.

3.4 Combined Effects of Impurities and Pinning

One should however consider the combined effects of the scattering off impurities and off vortices, induced by an external magnetic field, on the quasiparticle spectrum. The disorder was modelled [103] using the binary alloy model [22] where the impurities are distributed randomly over the lattice sites. At each impurity site it costs an energy U to place an electron (it acts as a local shift on the chemical potential). The impurities are randomly distributed over a $L \times L$ periodic two-dimensional lattice and play the role of pinning centers for the vortices. It is favorable that a vortex is located in the vicinity of an impurity [93]. Taking into account a given distribution of the positions of the impurities $\{\mathbf{r}_i^p\}_{i=1,N_p}$, the distribution of the positions of the vortices $\{\mathbf{r}_i^v\}_{i=1,N_v}$ is chosen in such a way it minimizes the total vortex energy given by $\mathcal{E} = \mathcal{E}_v + \mathcal{E}_p$, where $\mathcal{E}_v = \mathcal{U}_v \sum_{\mathbf{r}_i^v, \mathbf{r}_j^v} K_0 \left(\left| \mathbf{r}_i^v - \mathbf{r}_j^v \right| / \lambda \right)$ is the repulsive interaction energy between the vortices in the London regime. and $\mathcal{E}_p = \mathcal{U}_p \sum_{\mathbf{r}_i^v, \mathbf{r}_j^p} f \left(\left| \mathbf{r}_i^v - \mathbf{r}_j^p \right| / r_p \right)$ is the pinning energy associated with the impurities, acting as pinning centers for the vortices. The interaction between the vortices is not significantly screened since the penetration length is very large. In the equations above $\mathcal{U}_v = (\phi_0 / 4\pi\lambda)^2$ is the energy of interaction between two vortices, $K_0(r)$ stands for the zero order Hankel function, $\mathcal{U}_p < 0$ is the pinning strength created by an impurity and $f(r/r_p)$ is a rapidly decreasing function for $r/r_p > 1$. In the model it is assumed that the pinning energy is much larger than the interaction between vortices $|\mathcal{U}_p| \gg \mathcal{U}_v$ and $r_p \sim \delta$ where δ is the lattice constant. In that case, as $N_v \ll N_p$ each vortex is preferably pinned in the close

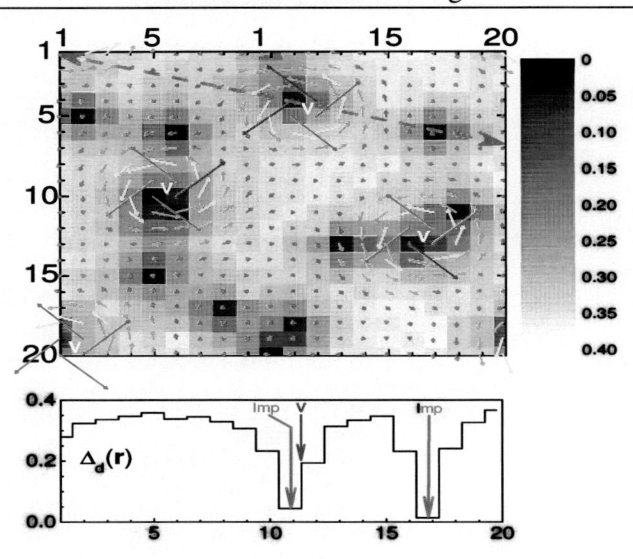

Figure 12: Dependence of the gap function on the positions of vortices and impurities. The darker areas correspond to the locations of the impurities.

vicinity of an impurity. As we take the London limit, which is valid for low magnetic field and over most of the $H - T$ phase diagram in extreme type-II superconductors such as cuprates, we assume that the size of the vortex core is negligible and place each vortex core at the center of a plaquette. So in the limit of strong pinning described above, each vortex will be pinned in the center of one of the four plaquettes surrounding a site hosting an impurity. The plaquetes selected by the vortices are those minimizing the interaction energy $\mathcal{E}_v$ between the vortices. Once the impurity positions $\{\mathbf{r}_i^p\}_{i=1,N_p}$ are fixed and the correlated vortex positions $\{\mathbf{r}_i^v\}_{i=1,N_v}$ are determined the quasiparticle spectrum is obtained solving the BdG equations $\mathcal{H}(\mathbf{r})\Psi_n(\mathbf{r}) = \epsilon_n\Psi_n(\mathbf{r})$, where $\Psi_n^\dagger(\mathbf{r}) = (u_n^*(\mathbf{r}), v_n^*(\mathbf{r}))$

The disorder potential induced by the impurities and the inhomogeneous superfluid velocities induced by the vortices strongly affect the pairing potential $\Delta(\mathbf{r}, \mathbf{r}+\delta)$ defined over the link $[\mathbf{r}, \mathbf{r}+\delta]$ of the two-dimensional lattice. Thus, for a given configuration of the impurity positions and of the vortex positions, the pairing potential $\Delta(\mathbf{r}, \mathbf{r}+\delta)$ is calculated self-consistently until convergence is obtained for each individual link. On each lattice site we can define the amplitude of the d-wave order parameter as $\Delta_d(\mathbf{r}) = \sum_\delta (-1)^{\delta_y}\Delta(\mathbf{r}, \mathbf{r}+\delta)$. The amplitude of the order parameter is strongly suppressed in the vicinity of impurities and vortices as seen in the Fig. 12.

In Fig. 13 we show the density of states averaged over 100 configurations for a moderate impurity concentration and for a typical value of $U = 5t$. We compare the cases with no vortices and with a low vortex density (considering both an energetically favorable distribution of the vortices and a fully random configuration). The self-consistent calculation of the order parameter gives a maximum amplitude, Δ_0, of the order of $0.5t$. When there are no vortices present, coherence peaks appear at $\epsilon \sim \Delta_0$. If the vortices penetrate the sample these peaks disappear. Without vortices the density of states vanishes at zero energy as found before and if we include the magnetic field the density of states becomes finite [21]. Also, it is clear that if the vortices are fully randomly distributed the DOS is

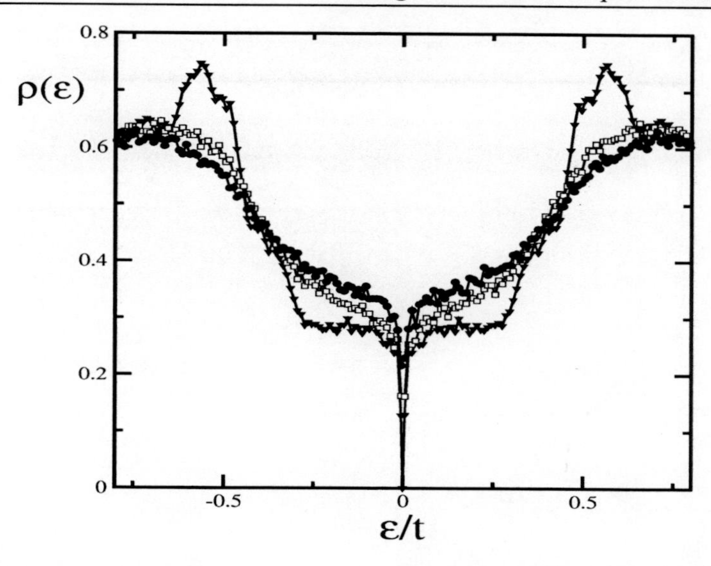

Figure 13: Quasiparticle density of states. We consider three cases: no vortices (filled triangle down), 4 vortices pinned at locations ensuring the minimization of the total vortex energy (square), and 4 vortices located at random (bullet).

larger than the one where the vortices tend to be pinned at the impurity sites. The results are therefore qualitatively similar to the ones obtained when there are no impurities present ($U = 0$) [21].

In addition it was shown [103] that the influence of the impurity concentration and of the repulsive local potential, U leads to no qualitative difference except that there is a slight increase in the DOS. Both sets of results indicate that the strong effect is the scattering off the vortices.

More detailed information about the scattering of the quasiparticles is obtained calculating the LDOS for a given impurity/vortex configuration. In Fig. 14 we compare the LDOS at four different sites [103]. i) At a site in the bulk the band-structure profiles are somewhat similar to the case of a vortex lattice [21]. The coherence peaks are evident, the zero energy density of states is very small (but not strictly zero), and the low energy peaks are smeared out. ii) At an impurity site (and no vortex nearby) the same structure is apparent except that since the impurity potential is repulsive the density of states at the impurity site is considerably depleted at low energy (for instance in the very large U limit the density of states is virtually zero at the impurity site). iii) In the vicinity of a vortex site (but far from any impurity) the structure is very similar to the case obtained before [21] with no coherence peaks close to the vortex and an enhanced zero energy density of states near the vortex. iv) Finally the interesting case of a location where a vortex is bound to an impurity reveals that the coherence peaks are recovered very close to the vortex. Also, the low energy density of states at the impurity site is increased with respect to the (Imp) case, due to the vortex nearby. However the density of states is considerably smaller than for the case (Vx) of a vortex far from any impurity. These results are in qualitative agreement with experimental results if most vortices are pinned at impurity sites. However, there is a small zero energy density of states in disagreement with the experimental results.

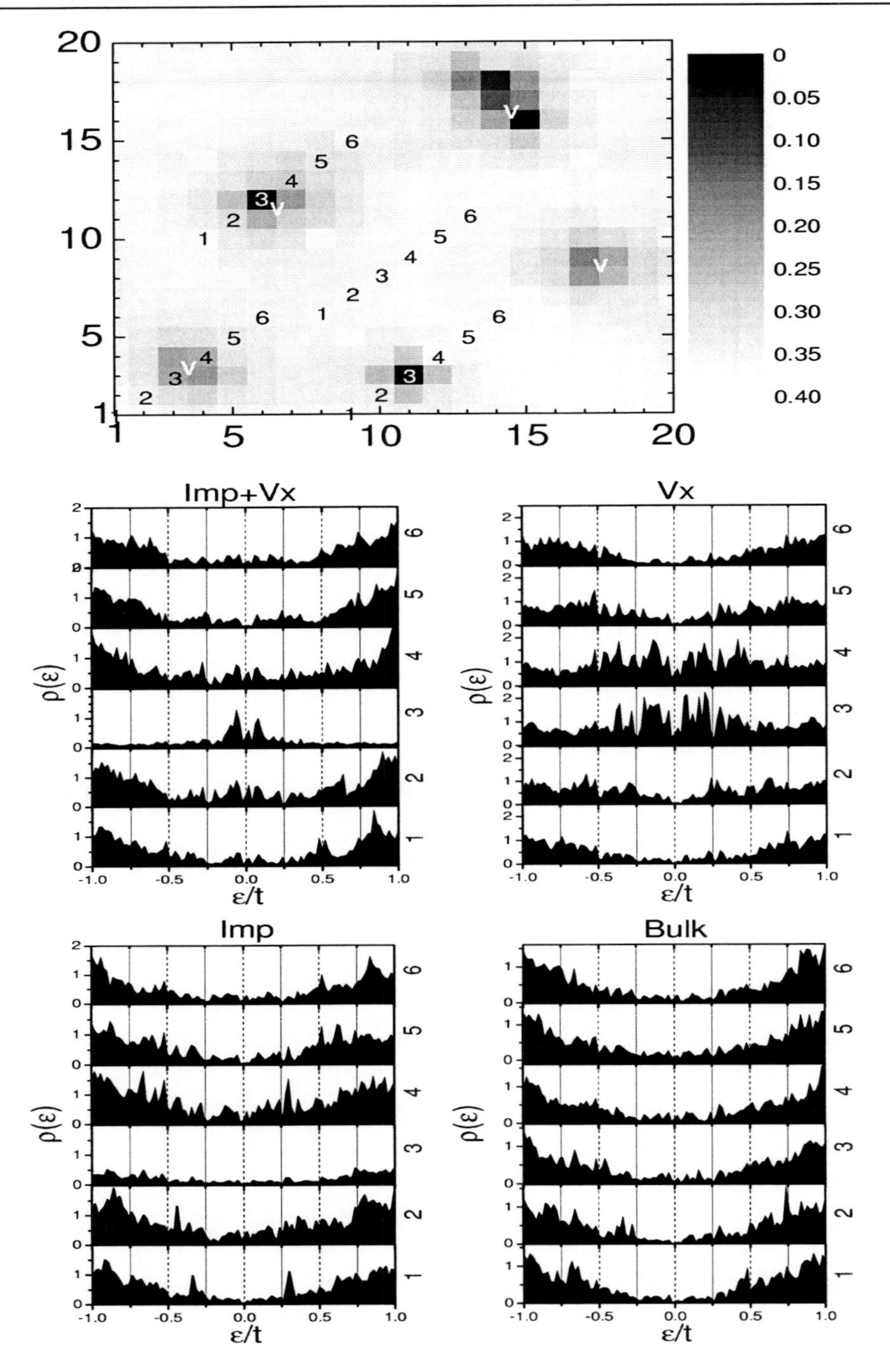

Figure 14: Quasiparticle local density of states for different points on the system (reproduced from [103]).

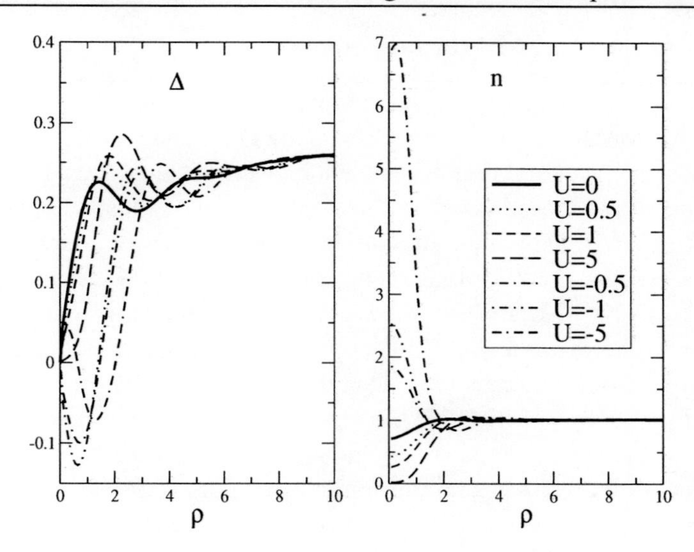

Figure 15: Δ and electron density as a function of distance for different values of U.

3.5 Effects of Disorder on the Vortex Charge

Consider now the effect of an impurity in a s-wave vortex. The Bogoliubov-de Gennes equations can be solved on a continuum [27] introducing an impurity as a disc of small radius centered at the vortex location. The results are presented in Fig. 15. From this figure we see that both the effect of temperature and of the impurity potential is significant. The effect of the impurity potential is to increase the positive charge if $U > 0$ and to change the sign, by attracting electrons to the vortex core, if $U < 0$, as expected. Also, a small negative U has an important effect on the charge. The effect of both temperature and positive U is to enlarge the normal core region (Kramer-Pesch effect). In the case of the attractive potential the gap function acquires a node near the origin and the charge accumulation near the origin is quite sharp for a moderate $|U|$ value.

Due to the random nature of the contribution of impurity potentials in a real disordered material (like in the Anderson model for localization), their effect on the distribution of vortex charges throughout the system is complex. Therefore, in general it is difficult to predict the outcome of their effect.

Considering positional disorder of the vortices without the presence of impurities for both s- and d-wave symmetries at intermediate values of the magnetic field the effect is very small, except when two or more vortices are close by, in which case the vortex charge may change sign, particularly in the s-wave case [94]. When impurities are added to the positional disorder they have a noticeable effect on the vortex charge. The effect of random impurities in the case of several vortices distributed in a lattice (dirty homogeneous superconductor) is very similar to the case of a single impurity [94]. This together with the effect of temperature changes appreciably the vortex charge. In the case of an attractive impurity potential the sign of the charge changes as for the single vortex case.

3.6 DOS, LDOS and STM Measurements

The results for the DOS and LDOS were compared to the experimental results for several systems, when the overlap of the vortices affects the DOS and when there is disorder in the system. In some systems other order parameters may compete or coexist with superconductivity, like in high-T_c superconductors and heavy fermion systems, and one has to take into account their effect on the DOS. In particular, in the case of d-wave symmetry it is predicted that in the vortex core local antiferromagnetic correlations may occur and as a consequence the quasiparticles may also scatter from these excitations in a non-trivial manner.

STM studies of the material $Bi_2Sr_2CaCu_2O_{8+\delta}$ revealed a modulation of the LDOS with a periodicity of four unit cells [95]. Due to the discovery of spin density modulations in the material $La_{1.84}Sr_{0.16}CuO_4$ [96] the modulations were interpreted as resulting from the coexistence of the spin order with superconductivity and the consequent scattering of the quasiparticles off the vortices and the spin modulation. Later studies related the quasiparticle interference and photoemission spectroscopy data and concluded that actually it is consistent to explain these LDOS modulations considering only the scattering of the quasiparticles from the vortex cores [97] and impurities. In particular, the Fourier transform of the LDOS was studied as a function of energy and interesting interference patterns were found. These were qualitatively explained considering the scattering of the quasiparticles off a single impurity [98].

Using a similar model and considering scattering over weak random impurities, it was concluded that the interference patterns can be used as a test for the type of dominant fluctuations in the system [99]. In particular, above the critical temperature and in zero magnetic field, the nature of the pseudogap phase of the cuprate superconductors has risen controversy. In one view the phase originates from the superconducting fluctuations where the gap amplitude has become finite but there is no phase coherence, and in other views the pseudo-order has its origin on other types of fluctuations (like charge d-wave). It was then shown [99] that the interference patterns depend in a distinctive and characteristic way on the coherence factors, enabling to pinpoint the dominant fluctuations as having a superconducting origin or not. The results await confirmation.

Note that other experiments for the material $Bi_2Sr_2CaCu_2O_{8+\delta}$ done in zero field have yielded a periodic density of states modulation, further fueling the discussion of the mechanism for the quasiparticle interference effects [100]. Also, as stressed in Ref. [101] one can not rule out the presence of the spin collective modes which may give a significant contribution to the STM data. A similar proposal was also presented in Ref. [102]. This remains an open issue.

4 Conclusions

In this chapter we covered different aspects of the influence of magnetic field and disorder in a strongly type-II superconductor. We payed attention to various regimes from a single vortex, through intermediate fields up to the high field regime where a Landau level description is appropriate.

We stressed the importance of the combined effects of the standard Doppler shift effect on the quasiparticles energies and the quantum Berry like term resulting from the scattering off the Aharonov-Bohm half-fluxes at the vortex cores. Also, we stressed the gapless nature of the spectrum in the high field limit and mentioned the possibility that the Hall conductance may be used as an order parameter characterising the transition in s-wave superconductors from the high field gapless regime to the low field gapped regime. In particular it was shown that in this regime the Hall conductance vanishes, as expected.

Next the density of states of a disordered superconductor in a pinned fully random vortex array was considered. Both the disorder and the magnetic field fill the density of states at low energies. In the s-wave case the density of states behaves as a power law with an exponent that scales with $1/\sqrt{B}$. In general one finds a finite density of states at zero energy for the d-wave case except in the limit of very small magnetic fields. The density of states deviates from the zero energy value by a power law. The zero energy density of states scales with the inverse of the magnetic length ($\sqrt{B}$). In the d-wave case the Dirac anisotropy further increases the weight of the density of states at low energies. Also it affects the exponent of the power law. In the isotropic case the exponent is 1 at very low energies and around 2 neglecting this narrow region. In the linear regime the density of states is of the form of a scaling function of the energy and the magnetic field. As the anisotropy increases this narrow regime shrinks considerably and, neglecting this region, the exponent of the density of states interpolates to 1, which is the Dirac limit. This limit is also obtained in the zero field limit in the isotropic case. Except for the zero energy finite value the energy dependence of the density of states in the case with disorder is similar to the lattice case. This suggests that the vortex disorder does not dramatically affect the density of states at low energies. An analysis of the IPR and the LDOS shows that the lowest lying states are delocalized, even though strongly peaked at the vortex cores. In the gapped s-wave case however the disorder introduces states in the gap thereby changing qualitatively the low energy density of states. A power law behavior with an exponent that scales with the magnetic length was found.

The results show that the dominant effect on the quasiparticle DOS is due to the vortex scattering. The presence of an impurity basically renormalizes the DOS except when the impurity is strongly repulsive. In this case the density of states is significantly depressed near the impurity. This is the unitary limit where a gapped system is predicted in the absence of magnetic field [13]. The quantum effect originated in the Aharonov-Bohm scattering of the quasiparticles circulating around a vortex line has been shown to have considerable effects on the density of states [67, 104]. Significant changes with respect to the classical Doppler shift effect [64] occur at low energies [67]. The results obtained show that the addition of impurity scattering is not very significant and the Berry phase is dominant, as argued before [29, 21]. The results are very similar to the ones obtained experimentally with STM in ref. [78], except for the finite density of states at zero energy in the vicinity of an isolated vortex. The experimental results are therefore more consistent with the situation where all or most of the vortices are pinned to the impurity sites. At these points, even though there is an enhancement of the low energy density of states with respect to an impurity with no vortex attached, the increase is reduced by the presence of the impurity with respect to the case of a vortex but no impurity nearby.

Also, we looked at the influence of disorder on the vortex charge both for a single vortex and a regime where several vortices are present. This may be relevant for instance on the context of the Hall anomaly and in general to the study of the electrostatics of the vortices.

Recently considerable attention has been devoted to the effect of competing order parameters on the DOS and LDOS of these systems. According to several theories this competition is unavoidable, at least in the vortex cores. The situation is however complex. The effect of competing order parameters has also been addressed in the context of the vortex charge. For instance the influence of antiferromagnetic fluctuations on the vortex charge have yielded results where the sign of the vortex charge becomes negative [39]. This effect disappears increasing the temperature such that the local antiferromagnetic order in the vortex cores vanishes [44].

The authors acknowledge many discussions and a fruitful collaboration over the years with Zlatko Tesanovic, and discussions with Sasa Dukan, Alexander Nersesyan, José Rodriguez, Pedro Bicudo and Marco Cardoso.

References

[1] A.A. Abrikosov, *Sov. Phys. JETP* **5**, 1174 (1957).

[2] M. Rasolt and Z. Tesanovic, *Rev. Mod. Phys.* **64**, 709 (1992).

[3] S. Dukan, A.V. Andreev and Z. Tesanovic, *Physica C* **183**, 355 (1991).

[4] S. Dukan and Z. Tesanovic, *Phys. Rev. B* **49**, 13017 (1994).

[5] M.R. Norman, A.H. MacDonald and H. Akera, *Phys. Rev. B* **51**, 5927 (1995).

[6] C. Caroli, P.G. de Gennes and J. Matricon, *Phys. Lett.* **9**, 307 (1964).

[7] P.W. Anderson, *Phys. Rev. Lett.* **3**, 325 (1959).

[8] K. Ueda and T.M. Rice, in *Theory of Heavy Fermions and Valence Fluctuations*, Ed. T. Kasuya and T. Saso (Springer, Berlin, 1985).

[9] P.A. Lee, *Phys. Rev. Lett.* **71**, 1887 (1993); Y. Hatsugai and P.A. Lee, *Phys. Rev. B* **48**, 4204 (1993).

[10] M. Franz, C. Kallin and A. J. Berlinsky, *Phys. Rev. B* **54**, R6897 (1996); M. Franz et al. *ibid* **56**, 7882 (1997).

[11] S. Haas, A.V. Balatsky, M. Sigrist and T.M. Rice, *Phys. Rev. B* **56** 5108 (1997).

[12] A.A. Abrikosov and L.P. Gorkov, *Sov. Phys. JETP* **12**, 1243 (1961).

[13] A. Altland, B. D. Simons, and M. R. Zirnbauer, *Phys. Rep.* **359**, 283 (2002).

[14] A. A. Nersesyan, A. M. Tsvelik, and F. Wenger, *Phys. Rev. Lett.* **72**, 2628 (1994); *Nucl. Phys. B* **438**, 561 (1995).

[15] T. Senthil, M. P. A. Fisher, L. Balents, and C. Nayak, *Phys. Rev. Lett.* **81**, 4704 (1998).

[16] R. B. Laughlin, *Phys. Rev. Lett.* **80**, 5188 (1998); T. Senthil, J. B. Marston, and M. P. A. Fisher, *Phys. Rev. B* **60**, 4245 (1999).

[17] S. Dukan and Z. Tešanović, *Phys. Rev. B* **56**, 838 (1997).

[18] P. D. Sacramento, *Phys. Rev. B* **59**, 8436 (1999).

[19] J. Ye, *Phys. Rev. Lett.* **86**, 316 (2001).

[20] D. V. Khveshchenko and A. G. Yashenkin, *Phys. Rev. B* **67**, 052502 (2003).

[21] J. Lages, P. D. Sacramento and Z. Tešanović, *Phys. Rev. B* **69**, 094503 (2004).

[22] W. A. Atkinson, P. J. Hirschfeld and A. H. MacDonald, *Phys. Rev. Lett.* **85**, 3922 (2000).

[23] A. Ghosal, M. Randeria and N. Trivedi, *Phys. Rev. B* **63**, 020505 (2000).

[24] E. R. Ulm et al., *Phys. Rev. B* **51**, 9193 (1995); D. N. Basov et al., *Phys. Rev. B* **49**, 12165 (1994); C. Bernhard et al., *Phys. Rev. Lett.* **77**, 2304 (1996); S. H. Moffatt, R. A. Hughes, and J. S. Preston, *Phys. Rev. B* **55**, 14741 (1997).

[25] A. Ghosal, M. Randeria, and N. Trivedi, *Phys. Rev. Lett.* **81**, 3940 (1998).

[26] P.G. de Gennes, *Superconductivity of Metals and Alloys* (Addison-Wesley, Reading, MA, 1989).

[27] F. Gygi and M. Schluter, *Phys. Rev. B* **43**, 7609 (1991).

[28] N. Hayashi, T. Isoshima, M. Ichioka and K. Machida, *Phys. Rev. Lett.* **80**, 2921 (1998).

[29] M. Franz and Z. Tesanovic, *Phys. Rev. Lett.* **80**, 4763 (1998).

[30] P. I. Soininen, C. Kallin, and A. J. Berlinsky, *Phys. Rev. B* **50**, 13883 (1994).

[31] D.I. Khomskii and A. Freimuth, *Phys. Rev. Lett.* **75**, 1384 (1995).

[32] G. Blatter, M. Feigel'man, V. Geshkenbein, A. Larkin and A. van Otterlo, *Phys. Rev. Lett.* **77**, 566 (1996).

[33] K. Kumagai, K. Nozaki and Y. Matsuda, *Phys. Rev. B* **63**, 144502 (2001).

[34] N. Hayashi, M. Ichioka and K. Machida, *J. Phys. Soc. Jpn.* **67**, 3368 (1998).

[35] M. Machida and T. Koyama, *Phys. Rev. Lett.* **90**, 077003 (2003).

[36] J. Goryo, *Phys. Rev. B* **61**, 4222 (2000).

[37] J. Kolacek, P. Lipavsky and E. H. Brandt, *Phys. Rev. Lett.* **86**, 312 (2001).

[38] K. Yasui and T. Kita, *Phys. Rev. Lett.* **83**, 4168 (1999).

[39] Y. Chen, Z.D. Wang, J.-X. Zhu and C.S. Ting, *Phys. Rev. Lett.* **89**, 217001 (2002).

[40] A.T. Dorsey et al., *Phys. Rev. B* **46**, 8376 (1992).

[41] D.P. Arovas, A.J. Berlinsky, C. Kallin and S.-C. Zhang, *Phys. Rev. Lett.* **79**, 2871 (1997).

[42] S. Chakravarty, R. B. Laughlin, D.K. Morr and C. Nayak, *Phys. Rev. B* **63**, 094503 (2001).

[43] M. Franz, D.E. Sheehy and Z. Tesanovic, *Phys. Rev. Lett.* **88**, 257005 (2002).

[44] Y. Chen, Z.D. Wang and C.S. Ting, *Phys. Rev. B* **67**, 220501 (2003).

[45] M.M. Maska and M. Mierzejewski, *Phys. Rev. B* **68**, 024513 (2003).

[46] D. Knapp, C. Kallin, A. Ghosal and S. Mansour, *Phys. Rev. B* **71**, 064504 (2005).

[47] R. Corcoran *et al.*, *Phys. Rev. Lett.* **72**, 701 (1994); G. Goll *et al.*, *Phys. Rev. B* **53**, 8871 (1996).

[48] S. Dukan and Z. Tešanović, *Phys. Rev. Lett.* **74**, 2311 (1995).

[49] P. Miller and B. L. Györffy, *J. Phys. Cond. Matter* **7**, 5579 (1995).

[50] G. M. Bruun, V. N. Nicopoulos, and N. F. Johnson, *Phys. Rev. B* **56**, 809 (1997); V. N. Nicopoulos and P. Kumar, *Phys. Rev. B* **44**, 12080 (1991).

[51] Z. Tesanovic and P.D. Sacramento, *Phys. Rev. Lett.* **80**, 1521 (1998).

[52] Norman M. R. and MacDonald A. H., *Phys. Rev. B* **54** 4239 (1996).

[53] A. L. Carr, J. J. Trafton, S. Dukan and Z. Tesanovic, *Phys. Rev. B* **68**, 174519 (2003).

[54] M. Nohara, M. Isshiki, F. Sakai and H. Takagi, *J. Phys. Soc. Jpn.* **68**, 1078 (1999).

[55] S. Dukan, T. P. Powell and Z. Tesanovic, *Phys. Rev. B* **66**, 014517 (2002).

[56] E. Boaknin, R. W. Hill, C. Proust, L. Taillefer and P. C. Canfield, *Phys. Rev. Lett.* **87**, 237001 (2001); E. Boaknin, M. A. Tanatar, J. Paglione, D. Hawthorn, F. Ronning, R. W. Hill, M. Sutherland, L. Taillefer, J. Sonnier, S. M. Hayden and J. W. Brill, *Phys. Rev. Lett.* **90**, 117003 (2003).

[57] T. Maniv, V. Zhuravlev, I. Vagner and P. Wyder, *Rev. Mod. Phys.* **73**, 867 (2001).

[58] J. E. Graebner and M. Robbins, *Phys. Rev. Lett.* **36**, 422 (1976).

[59] Y. Onouki, I. Umehara, T. Ebihara, N. Nagai and T. Takita, *J. Phys. Soc. Jpn.* **61**, 692 (1992).

[60] F. Mueller, D. H. Lowndes, Y. K. Chang, A. J. Arko and R. S. List, *Phys. Rev. Lett.* **68**, 3928 (1992).

[61] G. Kido, K. Komorita, H. Katayama-Yoshida and T. Takahashi, *J. Phys. Chem. Solids* **52**, 1465 (1991).

[62] N. Harrison, S. M. Hayden, P. Messon, M. Springford, P. J. van der Wel and A. A. Manovsky, *Phys. Rev. B* **50**, 4208 (1994).

[63] T. J. B. M. Janssen and M. Springford, in *The Superconducting State in Magnetic Fields*, ed. C. A. R. Sá de Melo, World Scientific (1998).

[64] G. E. Volovik, *Pis'ma Zh. Éksp. Teor. Fiz.* **58**, 457 (1993); [*JETP Lett.* **58**, 469 (1993)].

[65] Y. Aharonov and A. Casher, *Phys. Rev. Lett.* **53**, 319 (1984).

[66] M. Franz and Z. Tešanović, *Phys. Rev. Lett.* **84**, 554 (2000).

[67] O. Vafek, A. Melikyan, M. Franz, and Z. Tešanović, *Phys. Rev. B* **63**, 134509 (2001).

[68] M. Chiao, R. W. Hill, C. Lupien, B. Popić, R. Gagnon, and L. Taillefer, *Phys. Rev. Lett.* **82**, 2943 (1999).

[69] J. Lowell and J. B. Sousa, *J. Low Temp. Phys.* **3**, 65 (1970).

[70] S. H. Simon and P. A. Lee, *Phys. Rev. Lett.* **78**, 1548 (1997).

[71] L. P. Gor'kov and J. R. Schrieffer, *Phys. Rev. Lett.* **80**, 3360 (1998).

[72] A. S. Mel'nikov, *J. Phys. Cond. Matt.* **11**, 4219 (1999).

[73] H. F. Hess, R. B. Robinson, R. C. Dynes, J. M. Valles, Jr., and J. V. Waszczak. *Phys. Rev. Lett.* **62**, 214 (1989).

[74] H. F. Hess, R. B. Robinson, and J. V. Waszczak. *Phys. Rev. Lett.* **64**, 2711 (1990).

[75] M. Ichioka, N. Hayashi and K. Machida, *Phys. Rev. B* **55**, 6565 (1997).

[76] Y.-D. Zhu, F. C. Zhang and M. Sigrist, *Phys. Rev. B* **51**, 1105 (1995).

[77] I. Maggio-Aprile, Ch. Renner, A. Erb, E. Walker, and Ø. Fischer, *Phys. Rev. Lett.* **75**, 2754 (1995).

[78] S. H. Pan et al., *Phys. Rev. Lett.* **85**, 1536 (2000).

[79] Matsuda Y. *et al.*, *Phys. Rev. B* **52** R15749 (1995); Nagaoka T. *et al.*, *Phys. Rev. Lett.* **80** 3594 (1998).

[80] G. D'Anna, V. Berseth, L. Forró, A. Erb and E. Walker, *Phys. Rev. Lett.* **81**, 2530 (1998).

[81] Stephen M. J., *Phys. Rev. B* **43** 1212 (1991).

[82] Thouless D. J., *J. Phys. C* **17** L325 (1984).

[83] P. D. Sacramento, *J. Phys. Cond. Matt.* **11**, 4861 (1999).

[84] T. Senthil, J. B. Marston and M. P. A. Fisher, *Phys. Rev. B* **60**, 4245 (1999).

[85] O. Vafek, A. Melikyan and Z. Tesanovic, *Phys. Rev. B* **64**, 224508 (2001).

[86] Z. Gedik and Z. Tesanovic, *Phys. Rev. B* **52**, 527 (1995).

[87] V.G. Kogan, *J. Low Temp. Phys.* **20**, 103 (1975).

[88] Y.A. Bychkov and E.I. Rashba, *Sov. Phys. JETP* **58**, 1062 (1983).

[89] L. Marinelli, B. I Halperin, and S. H. Simon, *Phys. Rev. B* **62**, 3488 (2000).

[90] A. Vishwanath, *Phys. Rev. B* **66**, 064504 (2002).

[91] Y. Wang and A. H. MacDonald, *Phys. Rev. B* **52**, R3876 (1995).

[92] M. Chiao, R. W. Hill, C. Lupien, and L. Taillefer, *Phys. Rev. B* **62**, 3554 (2000).

[93] G. Blatter et al., *Rev. Mod. Phys.* **66**, 1125 (1994).

[94] J. Lages and P. D. Sacramento, *Phys. Rev. B* **73**, 134515 (2006).

[95] J. E. Hoffman, E. W. Hudson, K. M. Lang, V. Madhavan, H. Eisaki, S. Uchida and J. C. Davis, *Science* **295**, 466 (2002).

[96] B. Lake, G. Aeppli, K. N. Clausen, D. F. McMorrow, K. Lefmann, N. E. Hussey, N. Mangkorntong, M. Nohara, H. Takagi, T. E. Mason, A. Schröder, *Science* **291**, 1759 (2001).

[97] J. E. Hoffman, K. McElroy, D.-H. Lee, K. M. Lang, H. Eisaki, S. Uchida and J. C. Davis, *Science* **297**, 1148 (2002).

[98] Q.-H. Wang and D.-H. Lee, *Phys. Rev. B* **67**, R020511 (2003).

[99] T. Pereg-Barnea and M. Franz, *Phys. Rev. B* **68**, 180506 (2003).

[100] C. Howald, H. Eisaki, N. Kaneko, M. Greven and A. Kapitulnik, *Phys. Rev. B* **67**, 014533 (2003).

[101] A. Polkovnikov, S. Sachdev and M. Vojta, *Physica C* **388-389**, 19 (2003).

[102] C.-T. Chen and N.-C. Yeh, *Phys. Rev. B* **68**, R220505 (2003).

[103] J. Lages and P. D. Sacramento, *Phys. Rev. B* **71**, 132501 (2005).

[104] A.S. Mel'nikov, *Phys. Rev. Lett.* **86**, 4108 (2001).

In: Superconductivity, Magnetism and Magnets
Editor: Lannie K. Tran pp. 97-138 ISBN 1-59454-845-5

Chapter 4

SUPERCONDUCTING-FERROMAGNETIC HYBRID STRUCTURES AND THEIR POSSIBLE APPLICATIONS

Dimosthenis Stamopoulos*and Efthymios Manios
Institute of Materials Science, NCSR "Demokritos",
153-10, Aghia Paraskevi, Athens, Greece

Abstract

In recent years hybrid superconducting-ferromagnetic structures have attracted much interest not only due to their importance for basic physics but also for the numerous modern practical applications that could be based on such devices in the near future. In this work we present results on such hybrid structures (HSs) consisting of CoPt magnetic nanoparticles (MNs) that are randomly embedded at the bottom of high quality Nb superconducting layers. More specifically, in the present work we investigate the influence of the macroscopic magnetic characteristics of the MNs (i.e. saturation and coercive fields, saturation magnetization e.t.c.) on the nucleation of superconductivity in the Nb layer. The effects under discussion are studied by means of magnetization and resistivity measurements in various categories of HSs where MNs that exhibit different magnetic properties are employed. Our results demonstrate in a definite way that the MNs can be used at will as an efficient ingredient for the modulation of the superconducting properties of the Nb layer. While in a single superconductor the nucleation of superconductivity depends only on the applied magnetic field and temperature it turns out that in HSs the nucleation of superconductivity can be modulated not only by the parameters mentioned above but also by the magnetic state of the MNs. These facts are clearly revealed by the pronounced magnetic memory effects observed in our data. Furthermore, we find that in our HSs surface superconductivity is strongly enhanced in the high-field regime since on the operational $H-T$ phase diagram the respective boundary line $H_{c3}(T)$ exhibits a pronounced upturn when the applied field exceeds the saturation field of MNs. The observed change in slope of $H_{c3}(T)$ is determined by the saturation magnetization of MNs. Based on our experimental findings we propose possible practical applications that such HSs could find in the near future. Power applications, where the increasing of current-carrying capability of a superconductor is the main objective, and magnetoresistive memory elements,

*E-mail address: densta@ims.demokritos.gr. Author to whom correspondence should be addressed

where the element's resistance should ideally switch between two or three distinct states are discussed.

1 Introduction

Superconductivity [1, 2] and magnetism [3, 4] are probably the most widely studied collective phenomena from both theoretical and experimental point of view in solid state physics. When a type-II superconductor is cooled below its critical temperature T_c the application of an external magnetic field introduces magnetic vortices and depending on the Ginzburg-Landau parameter $\kappa = \lambda/\xi$ the so-called mixed state occupies a great part on the $H - T$ phase diagram. When the critical temperature T_c is exceeded superconductivity is destroyed due to the thermal breaking of Cooper pairs. In addition, superconductivity is also destroyed under the presence of an applied magnetic field even at $T = 0$.

Since an applied magnetic field destroys superconductivity there are serious limitations in the range of fields where a superconductor can be used for practical applications without exhibiting losses. A number of theoretical and experimental works have proposed possible candidate mechanisms to remove these limitations of tolerable magnetic fields. The most exotic proposal is the right combination of a superconductor with an appropriate magnetic material [5, 6, 7, 8]. The specific idea relies on doping a superconductor with appropriate magnetic impurities. The magnetic dopant should be chosen so that its coupling to the superconducting electron spins should be antiferromagnetic. In this way the mean field of the magnetic impurities will effectively compensate the external magnetic field so that the degeneracy between electrons of opposite spin and momentum will be restored [5, 6]. Consequently, the superconductor could preserve the desired superconducting properties at a specific range of external fields determined by the exchange field of the magnetic dopant.

Although this idea could be really effective when trying to enhance the range of magnetic fields where the superconductor preserves its main superconducting properties there are more problems that should be overcome in order to make the superconductor appropriate for practical applications. As discussed above in its mixed state a type-II superconductor hosts magnetic vortices that are related to normal state regions. When these vortices move under the application of a transport current Joule losses are introduced. Thus, in order to really improve the superconducting properties we should find a way to immobilize magnetic vortices. The introduction of structural defects (which may be uncorrelated as point defects or correlated as twin boundaries or of columnar type) in the superconductor by chemical substitution [9] or irradiation [10, 11, 12, 13] means may effectively lower the mobility of magnetic vortices. The critical currents obtained in this way exhibit satisfactory values. A second, more exotic way, to immobilize vortices is by introducing artificial patterns of *periodic* micro-sized or even submicro-sized magnetic dots which are embedded in the superconductor [14, 15, 16, 17, 18, 19, 20, 21, 22]. Advances in fabrication techniques have only very recently enabled the reliable preparation of such HSs [16, 17, 18, 19]. In this way we obtain enhanced critical currents only in a narrow range around some specific values of the external field where matching of vortices to the magnetic structure is achieved [14, 15, 16, 17, 18, 19, 20, 21, 22]. Furthermore, it was proved that the MNs impose on the superconductor the controlled modulation of superconducting order parameter. Thus such a HS may be considered us a system where the superconducting order parameter is *spatially modulated* under the influence of an *extrinsic artificial cause*. As discussed above it was observed that under certain conditions, depending mainly on the alignment of the MNs, the resistance in the HS maintained a lower value than the one observed in a reference pure superconducting layer. Most of the studies attributed the lowering of the measured resistance to the enhancement of the *bulk* and/or the *surface* critical current that the superconductor may sustain [16, 17, 18, 19, 23, 24, 25, 26]. This fact makes a HS an appropriate candidate for applications where increased current-carrying capability is the main objective. Except for current-carrying applications HSs could be also useful for the design of magnetoresistive memory devices or superconductive spin valves [27, 28, 29]. Very soon such devices could serve as elements for storage of data in a similar way to other candidate devices that are based on different physical mechanisms (for example giant magnetoresistance devices).

In this work we study how the superconducting properties of a low-T_c, type-II superconductor are modified under the influence of *randomly distributed* MNs which exhibit different magnetic characteristics. Nb and CoPt MNs were chosen as the ingredients of the HSs since their respective superconducting and magnetic properties are well studied and can be modified in a controlled way by altering the preparation conditions during sputtering and subsequent annealing [32, 23, 24, 25, 26, 30, 31]. More specifically, Nb was chosen since uncontested knowledge is established today as a result of extensive experimental studies that were performed in the last decades. As a consequence, a reliable comparative study could be performed between any HS and a reference pure Nb film. In our work by employing a specific procedure during the deposition process (see below) we achieved to prepare Nb layers of high quality as this is evidenced by their high $T_c = 8.4$ K, the sharpness of their superconducting transition $\Delta T = 50$ mK and finally the minor suppression of the superconducting critical temperature due to the magnetic constituent in the HS. The CoPt MNs employed in this work exhibit strong magnetic anisotropy with their easy-axis $\hat{a}_e$ of magnetization normal to the surface of the film. We concentrate our study in HSs consisting of randomly distributed MNs since their much easier preparation procedure (when compared to the advanced electron-beam lithography techniques needed for the production of ordered arrays) makes them more attractive for commercial applications [23, 24, 25, 26]. X-ray diffraction (XRD), scanning electron microscopy (SEM), transmission electron microscopy (TEM) and atomic force micorscopy (AFM) data revealed that the MNs are isolated particles with typical in-plane dimensions in the range $50 - 500$ nm and height approximately $10 - 50$ nm.

We investigate two categories of HSs. In the first category we studied HSs that consist of two-phase CoPt MNs that exhibit two different macroscopic coercive and saturation fields. This means that in these HSs we have two different kinds of MNs simultaneously embedded at the bottom of the same Nb film of thickness $d = 160$ nm. Due to their noticeable different coercive and saturation fields for the sake of clarity we call the MNs that exhibit the lower values as "soft" (SMNs) and the ones that exhibit the harder behaviour as "hard" (HMNs) in the rest of this chapter. In the second category of HSs the MNs are preferably embedded in only *half* of a high quality Nb layer of thickness $d = 200$ nm. In this way the modulation of the superconducting properties may be studied directly by performing resistance measurements on the hybrid part and the pure superconducting part of the combinatorial structure (CS) *simultaneously* under the application of the *same* dc transport current. The appearance of spatially modulated superconductivity and site-selective memory effects give direct evidence that in the HS part of the CS the nucleation of superconductivity can be greatly modulated by the dipolar fields of the MNs. The enhancement of the superconducting regime on the $H - T$ operational diagram suggests that such HSs could be attractive for current-carrying applications. In addition, a tristate superconducting magnetoresistive elemental device could be based on our CS.

In Section II we present the preparation procedure of the samples, detailed XRD data and the experimental details of the performed measurements. Both magnetic and magnetoresistance measurements have been performed. The obtained experimental data are presented in Sections III and IV for the first and second category of HSs respectively. A summary of our experimental results is given here. Both the bulk upper-critical $H_{c2}(T)$ and surface-superconductivity critical $H_{c3}(T)$ fields of the Nb host layer are affected by the presence of the MNs. Bulk superconductivity is slightly *promoted* by the SMNs, while it is definitely *depressed* by the HMNs. In contrast, surface superconductivity is enhanced under the influence of both SMNs and HMNs. This is evidenced by the fact that the respective line $H_{c3}(T)$ presents a clear upturn at the respective characteristic saturation field $H_{\rm sat}^{\rm MNs}$ of MNs. *As a result, in our HS the regime where the resistance is lower than the normal state value is greatly enhanced compared to a single Nb layer.* More specifically the slope of the upturn observed in $H_{c3}(T)$ is mainly determined by the saturation magnetization $m_{\rm sat}^{\rm MNs}$ of the MNs. Finally and more importantly, in our HSs the superconducting properties of the Nb layer exhibit *memory effects* in the regime above the bulk upper-critical field $H_{c2}(T)$. These *memory effects* are definitely not observed in a pure Nb layer and could be only attributed to the presence of MNs. In Section V a complete comparison is made between our experimental results and theoretical proposals presented in recent studies on relevant HSs. Our experimental findings are in nice qualitative agreement with the theoretical proposals of A. Yu. Aladyshkin et al. (Ref.[34]) and of Sa-Lin Cheng and H.A. Fertig (Ref.[35]). In Section VI a discussion is given on the possible applications that such HSs could find in the near future. Motivated by our experimental results we propose that the use of MNs with high magnetization and low saturation field could eventually lead to great broadening of the $H - T$ operational regime where a superconductor could be used for current-carrying applications. In addition, the memory effects observed in our HS could be important for the design of magnetoresistive memory storage elements [24, 25, 26]. Finally, in Section VIII the experimental findings and conclusions of the present chapter are summarized.

2 Preparation of the HS and Experimental Details

The films of Nb were sputtered from a $2''$ pure Nb target on Si $[001]$ substrates under an ultra pure Ar environment (99.999 %) of pressure 3 mTorr [32]. In order to find the appropriate power supplied to the DC gun for the deposition of Nb we examined in detail the influence of this parameter on the quality of the produced films. As a criterion, we measured the upper-critical field line $H_{c2}(T) = \Phi_o/2\pi\xi^2(T)$ of the deposited Nb films. The films that were produced at a DC power of 57 Watt (deposition rates of the order of 3.5 Å/sec) exhibited the smallest slope $dH_{c2}(T)/dT$. This indicates that these films exhibit the highest coherence length $\xi(0)$ that may be achieved in our case. Since the coherence length $\xi(T)$ is a parameter which is directly related to the electrons' mean-free path l and thus to the quality of the specimen we conclude that the DC power of 57 Watts is the optimum that

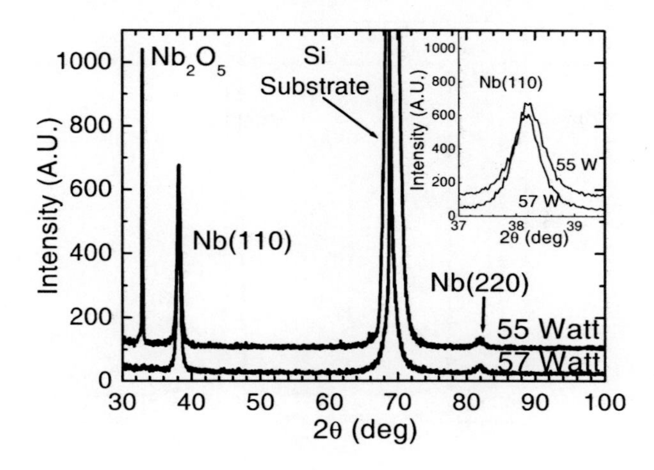

Figure 1: Detailed XRD data for two representative Nb films sputtered at 55 and 57 Watts. While the film produced at 55 Watts exhibits a pronounced peak referring to Nb_2O_5 the respective produced at 57 Watts exhibits only the (110) and (220) Nb peaks a fact indicating strong (110) texturing. In the inset we focus on the (110) peak of Nb.

may be used for the preparation of Nb films in our sputtering system. It should also be noted that the quality of the sputtered Nb films strongly depends on the residual oxygen existing in the chamber. In order to optimize this parameter we employed the following procedure. Since Nb is a strong absorber of oxygen, in cases when much residual oxygen existed in the chamber Nb oxide unavoidably was grown along the grain boundaries of Nb. In such cases the desired superconducting properties of Nb films are strongly depressed. In order to eliminate the residual oxygen that possibly existed in the chamber even after long periods $(2-3$ days) of pumping, we performed pre-sputtering for very long times. During the pre-sputtering all the residual oxygen was absorbed by the dummy deposited Nb. As a result we observed that after a pre-sputtering time of one hour the base pressure in the chamber was improved almost one order of magnitude exhibiting a typical value of 8×10^{-8} Torr (prior to pre-sputtering the base pressure in our chamber was of the order 10^{-7} Torr). Only when the desired base pressure was obtained we started the actual deposition of the Nb layers. This procedure had a strong consequence on the quality of the produced Nb films.

Detailed XRD for the produced Nb films are presented in figure 1 for two representative films produced at DC power of 57 and 55 Watts. In the inset we focus on the pronounced (110) peak of Nb in order to be clearly resolved for both films. As may be seen a small modification of the DC power supply (all other parameters are exactly the same) can result in the growth of Nb oxides. For film produced at 57 Watts only Nb peaks can be seen, but for the one produced at 55 Watts a pronounced peak appears for $2\vartheta = 32.9^o$, which refers to Nb_2O_5. The presence of Nb oxide in the specimen produced at 55 Watts is also reflected in the corresponding SQUID magnetic measurements since its critical temperature is strongly suppressed and the upper critical field line is placed to high magnetic fields when compared to the film produced at 57 Watts. At this point the crucial role of long pre-sputtering times should be stressed. If this step is omitted, the XRD patterns reveal the strong presence of Nb oxides. Both films presented in figure 1 have almost the same thickness, which is 1000

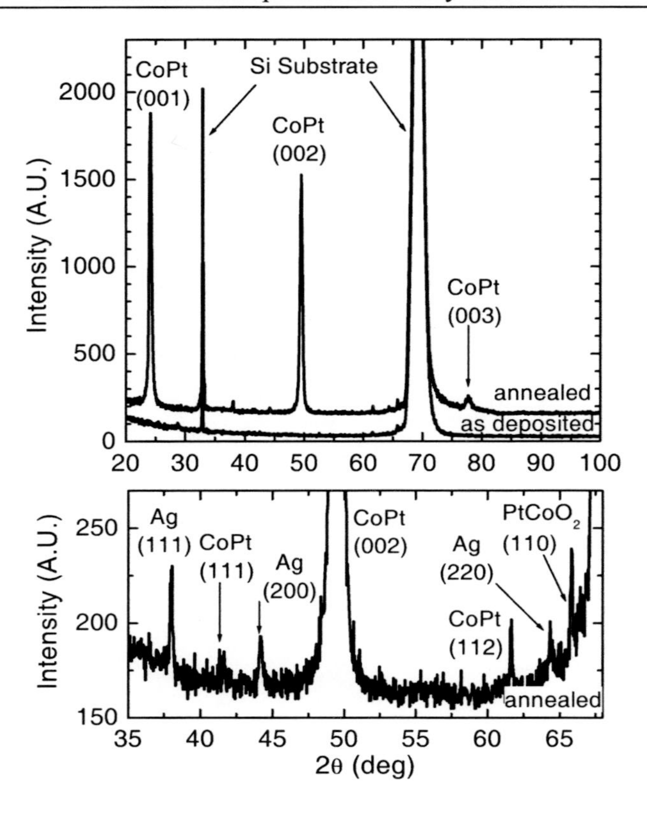

Figure 2: Detailed XRD data for Ag/CoPt sputtered films prior and after annealing. The pronounced (001), (002) and (003) peaks of CoPt, in contrast to its weak (111) and (112) peaks, is a strong indication of a high (001) texturing of the $L1o$ CoPt phase produced after annealing.

Å, and thus the same intensity for the strongest peak (110). Referring to the Nb specimen produced at 57 Watts we note that the absence of peaks other than (110) and (220) indicates (110) texture of the films used in the study presented in this chapter.

The MNs employed in our HSs were prepared as following. In order to obtain anisotropic and isolated CoPt nanoparticles a small quantity of Ag (nominal thickness of the deposited layer $d_{\rm Ag} = 0.3 - 1.2\,{\rm nm}$) should be used as an extra underlayer [25, 26, 30, 31]. The films were deposited on $\rm Si(001)$ substrates by magnetron sputtering at room temperature. The base pressure in our chamber was 5×10^{-7} Torr, while the pressure of Ar $(99,999\%$ pure) during sputtering was 3 mTorr. The nominal thickness of Ag layers for the MNs used in the HSs presented in this chapter is $d_{\rm Ag} = 0.6$ nm and were sputtered by using a DC power of 5 W on a $2''$ Ag target. The respective deposition rate was 1.5 Å/sec. The nominal thickness of CoPt layers $(50/50$ stoichiometry) is $d_{\rm CoPt} = 12$ nm and were sputtered by using an rf power of 30 W on a $2''$ CoPt target at a rate of 1.2 Å/sec. CoPt films deposited near room temperature adopt a disordered face centered cubic (fcc) structure which is magnetically soft. In order to form the hard magnetic and highly anisotropic $L1o$ phase (ordered face centered tetragonal or fct) of CoPt, the as deposited films need to be annealed under high vacuum $(10^{-7}$ Torr). When a single magnetic phase of CoPt is required both the annealing temperature and time should exceed a threshold value. For the Ag/CoPt bilayers the annealing temperature can't be reduced below $T = 600$ C while the respective time should exceed 20 min.

As it can be seen in the upper panel of figure 2, in the XRD data of the as-deposited Ag/CoPt films only the peaks of the Si substrate are distinguished from background. This is due to the small mean size of Ag and CoPt grains. During the annealing process CoPt is subjected to two kinds of transformations: (a) $L1o$ ordering and (b) recrystallization, grain growth and sintering [33]. As a result, after annealing there are grains of sufficient size to give strong peaks of $L1o$ CoPt. The most pronounced peak is the (001), the presence of which is the most obvious evidence for the formation of $L1o$ phase (this peak is forbidden for the disordered phase). The other peaks of CoPt that are clear are (002) and (003) (the (003) is also forbidden for disordered phase). In the lower panel of figure 2 we focus on the extremely weak (111) and (112) peaks of CoPt and on the (111), (200) and (220) peaks of Ag. In addition, a negligible amount of $PtCoO_2$ has been formed after annealing. The fact that $I_{001} > I_{002}$ indicates that the transformation of CoPt to the $L1o$ phase is almost complete. Nevertheless, due to its high saturation magnetization the small amount of the disordered (magnetically soft) phase of CoPt that remains when the annealing time is kept on purpose below 20 min is enough to give the broken loops observed in SQUID magnetization measurements (see figure 4(a) below). Also the fact that the only strong peaks of CoPt are (001), (002) and (003), while the (111) peak is weak, is a strong indication of a high degree of (001) texture and perpendicular magnetic anisotropy. This is in agreement with the results of the loop measurements presented below (see figures 4(a)-(b)).

As discussed above in the first category of HSs two-phase MNs are simultaneously embedded in a single Nb film. In order to produce this specific category of specimens we kept the temperature at $T = 600$ C and by fine adjustment of the annealing time at $15 - 18$ min we managed to obtain MNs that were *partially* transformed from their soft to the hard phase. In the second category the MNs exhibit a single magnetic phase and are preferably embedded in only half of a high-quality Nb layer. In both categories of studied HSs the produced MNs are anisotropic. Their easy-axis $\hat{a}_e$ of magnetization is normal to the surface of the film. More information on their magnetic characteristics is presented in the following Sections of this chapter. As already discussed, morphologically, the MNs are isolated and randomly distributed on the substrate's surface as may be seen in the SEM and AFM images presented in figure 3. The typical in-plane size and distance of the MNs are in the range $50 - 500$ nm. Both AFM data (presented) and cross-sectional TEM data (not shown) revealed that their mean thickness is of the order of $10 - 50$ nm. After producing the MNs, the Nb layer was sputtered on top of them according to the procedure outlined above.

The magnetoresistance measurements were performed by applying a dc transport current always normal to the magnetic field and measuring the voltage in the standard four-point pattern. A wide range of applied current was studied. In most of the measurements the applied current was $I_{\rm dc} = 0.5$ mA. Since the typical in-plane dimensions of the films are 1×4 (mm)^2 to 4×4 (mm)^2 and their thickness is in the range $100 - 200$ nm this value corresponds to an effective current density $J_{\rm dc} \approx 100 - 1000$ A/cm^2. The magnetic measurements were performed under both zero field cooling (ZFC) and field cooling (FC) conditions. The temperature control and the application of the dc fields were achieved in a

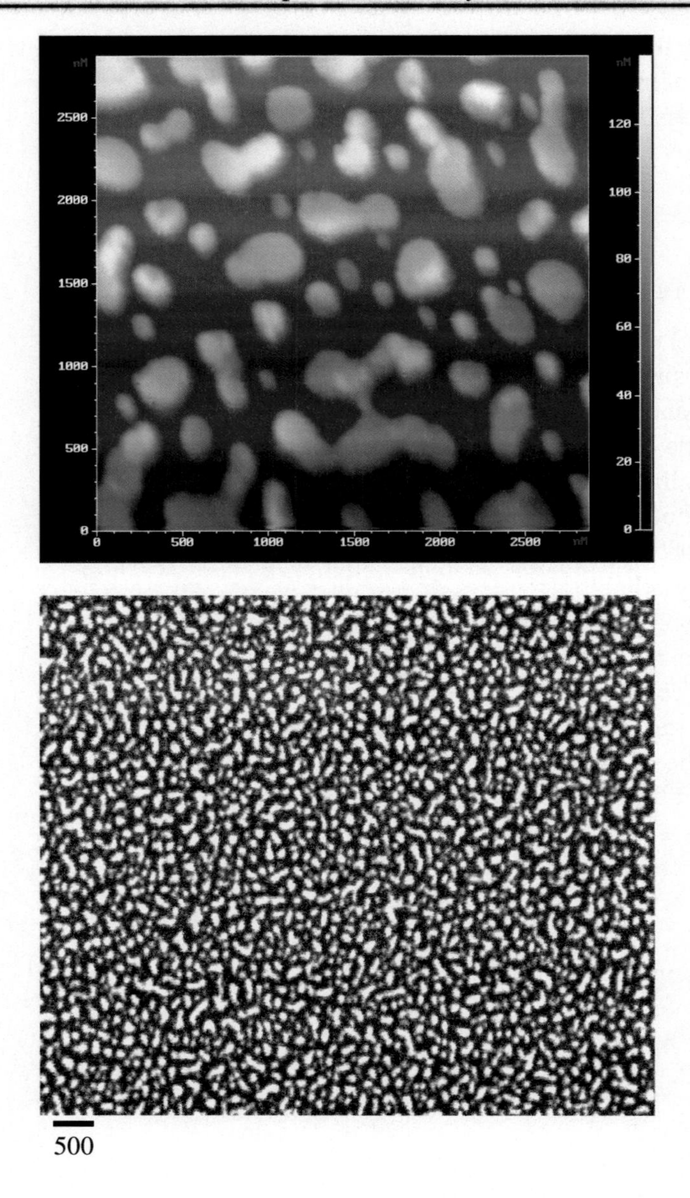

Figure 3: Images of the MNs by AFM (upper panel) and SEM (lower panel). The SEM image presents an extended area of the film, while the AFM one focuses on a much smaller regime. As we see the MNs are randomly distributed on the Si substrate and *isolated* from each other. Their typical in-plane dimension and distance are in the range $50 - 500$ nm. Their mean thickness is of the order of $10 - 50$ nm.

commercial SQUID device (Quantum Design). In all cases the applied field was parallel to the easy-axis $\hat{a}_e$ of magnetization ($\mathbf{H} \parallel \hat{a}_e$). We examined the whole temperature-magnetic-field regime accessible by our SQUID ($H_{\rm dc} < 55$ kOe, $T > 1.8$ K).

3 Experimental Results

3.1 First Category of HSs: Two-Phase MNs Embedded in a Single Nb Layer

In figures 4(a)-(b) we present the characteristic magnetization loops of the MNs at $T = 5$ K prior to the deposition of the Nb layer when the magnetic field is normal (upper panel) and parallel (lower panel) to the film's surface. We observe that the MNs are clearly anisotropic with the easy-axis $\hat{a}_e$ of magnetization normal to the surface of the film. By carefully regulating the annealing temperature and time of the primary magnetic layers we managed to get a mixture of both SMNs and HMNs on the same substrate. Since we wanted to study the influence of the saturation and coercive fields of MNs on the properties of a superconductor this was the only reliable way to exclude even slight sample-dependent variations of the superconducting properties which may be observed in different Nb samples. The saturation field of SMNs is $H_{\rm sat}^{\rm soft} \simeq 8$ kOe, while the HMNs exhibit $H_{\rm sat}^{\rm hard} \simeq 36$ kOe. The onset point where the contribution of the HMNs overcomes the saturation magnetization of the SMNs is $H_{\rm onset}^{\rm hard} \simeq 20$ kOe. Finally, the saturation magnetization of the HMNs is almost two times the one of SMNs, $m_{\rm sat}^{\rm hard} \simeq 2m_{\rm sat}^{\rm soft}$. It should be noted that the magnetization loop presented in figure 4 (a) don't change significantly in the low-temperature regime (2 K$< T < 10$ K) where the measurements have been performed, so that the characteristic saturation fields are practically constant in this temperature range. A representative measurement of the saturation magnetization in the low-T regime is presented in the inset of figure 4 (a).

After the complete magnetic and XRD characterization of the produced MNs we deposited the Nb layer according to the preparation procedure described in Section II. During the same deposition run a Si $[001]$ substrate was mounted next to the magnetic one in order to obtain a pure Nb film as a reference for our study. The critical temperatures of the produced HS and pure Nb are $T_c = 8.4$ K and $T_c = 8.6$ K respectively. The slight suppression of T_c in the HS indicates the high quality of the Nb layer deposited on the MNs. Figure 5 shows representative magnetic measurements as a function of temperature for magnetic fields $H_{\rm dc} = 5, 10$ and 15 kOe normal to the surface of the HS (full squares) and of pure Nb (open squares), while in the inset we present the normalized resistance curve (relatively to the normal state value just above the $T_c = 8.4$ K) of the HS from the superconducting state up to $T = 300$ K. In the magnetic data we present both curves for zero field cooling and subsequent field cooling. We observe that the bulk upper-critical temperatures $T_{c2}(H)$, defined as the point where the diamagnetic signal becomes zero almost coincide for the HS and pure Nb in this region of magnetic fields. Not surprisingly, the curves of the HS in the normal state exhibit a magnetic background which increases with the applied field. The background originates from the magnetization of the MNs. Despite that, the overall behaviour of the respective curves is almost the same in both the HS and pure Nb. This gives an additional indication that the quality of the Nb layer deposited on the MNs is almost the same with the one of the pure Nb film. The magnetically determined irreversibility points

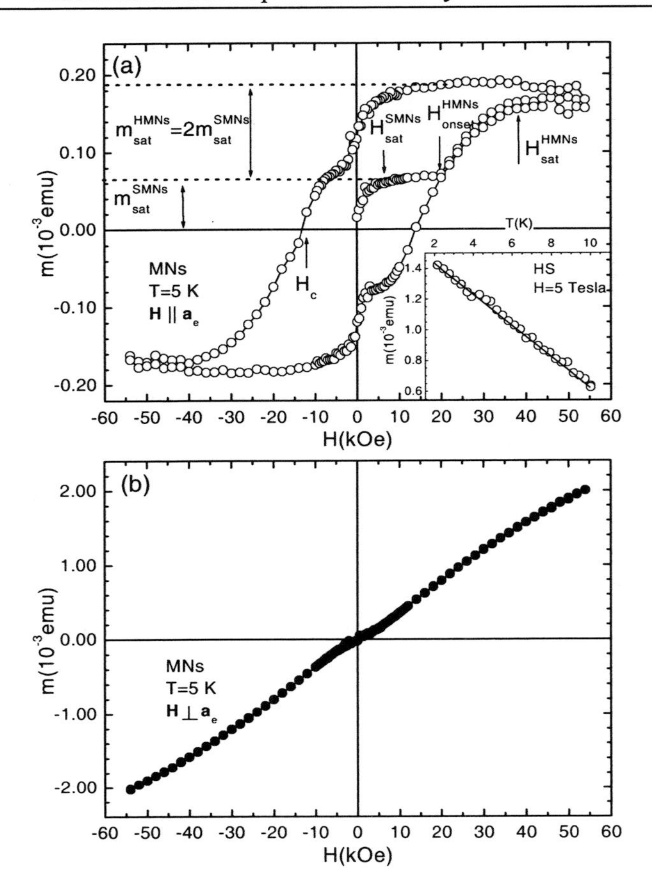

Figure 4: (Upper panel) Representative hysteresis loop for the MNs at $T = 5$ K when the magnetic field is normal (parallel) to the film (easy-axis $\hat{a}_e$ of magnetization). (Lower panel) Respective data for the case where the external field is parallel (normal) to the film (easy-axis $\hat{a}_e$ of magnetization). We clearly see that the anisotropic MNs exhibit both soft and hard magnetic behaviour. In the inset we present the low-temperature variation of the saturation magnetization under a magnetic field $H = 5$ Tesla for the case where $H \parallel \hat{a}_e$.

$T_{\rm irr}(H)$ are clearly placed at much lower temperatures compared to $T_{\rm c2}(H)$. This fact indicates that the identified upper-critical points $T_{\rm c2}(H)$ refer to the zeroing of the equilibrium magnetization and not to a collapse of screening currents that originate from bulk pinning.

In figure 6(a) we present a complete set of voltage measurements in the HS at a constant current density $j_{\rm dc} = 210$ A/cm^2 for various magnetic field's values. The configuration of the measurements is shown schematically in the inset of figure 6(b). We observe that at low magnetic fields the voltage curves are sharp but become very broad when high fields are applied. A comparison with the magnetization data revealed that the points where the voltage starts taking non-zero values almost coincide with the magnetically determined bulk upper-critical points $T_{\rm c2}^{\rm mag}(H)$. We clearly observe that especially in high magnetic fields the voltage curves take the normal state value at temperatures $T_{\rm c3}(H)$ much higher than $T_{\rm c2}^{\rm mag}(H)$ (see the curve referring to $H_{\rm dc} = 15$ kOe). In figure 6(b) we present the data

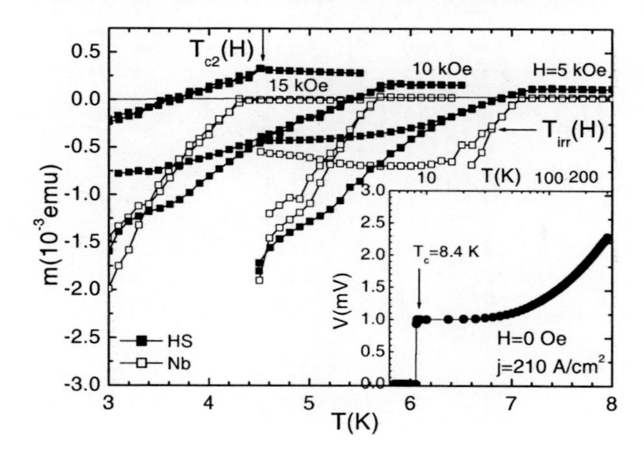

Figure 5: Representative magnetic data for both the HS (filled squares) and pure Nb (open squares) as a function of temperature for magnetic fields $H_{\rm dc} = 5, 10$ and 15 kOe parallel to the easy-axis $\hat{a}_e$ of magnetization ($\mathbf{H} \parallel \hat{a}_e$). The irreversibility points $T_{\rm irr}(H)$ are clearly distinct from the bulk upper-critical temperatures $T_{\rm c2}(H)$. In the inset we present in semilogarithmic plot the normalized voltage curve of the HS in an extended temperature regime. We observe that the residual resistance ratio is $RRR \approx 2.3$.

at high magnetic fields where superconductivity is gradually destroyed. More specifically, we see that for 36 kOe$< H <$ 46 kOe the characteristic point $T_{\rm c3}$ is almost constant and the voltage curves get broader as they tend to become horizontal. Above $H = 46$ kOe the measured curves present a distinct minimum at $T \approx 2.4$ K and their low-temperature part exhibits reentrance to the normal state. This surprising feature is reminiscent of (a) the Kondo effect where the resistance presents a minimum as a function of temperature and (b) the insulating behaviour which is usually observed in two dimensional superconducting or metallic films where the resistance increases logarithmically with decreasing temperature as a result of weak localization and electron-electron interaction [36]. We believe that in our case in the high-H low-T regime the HS reenters to the normal state rather than an insulating state. In order to study the dynamic behaviour in the specific high-H low-T regime under discussion we performed detailed $I-V$ characteristics. The data are presented in figure 6(c) in the form of dV/dI curves for $T = 2$ K. We see that in the whole range of high magnetic fields accessible in our SQUID the $I-V$ curves exhibit a linear behaviour (dV/dI curves are constant except for the curve referring to $H_{\rm dc} = 42$ kOe). Thus, we believe that the behaviour observed in our high-H low-T data presented in figure 6(b) is not related to a superconductor to insulator transition.

It should be noted that except for the high-H low-T regime discussed above, in the regime $T^{\rm mag}_{\rm c2}(H) < T < T_{\rm c3}(H)$ the resistance presents a clear non-linear behaviour. This fact is clearly indicated by the data presented in figures 7(a)-(c). In the upper (middle) panel we present normalized $V(T)$ data for $H = 20$ kOe and low (high) transport currents. We clearly observe that at low transport currents linear behaviour is exhibited in the regime close to $T_{\rm c3}(H)$ but when the applied current exceeds some critical value the non-linear character is clearly revealed. The linear behaviour is attained only when the characteristic

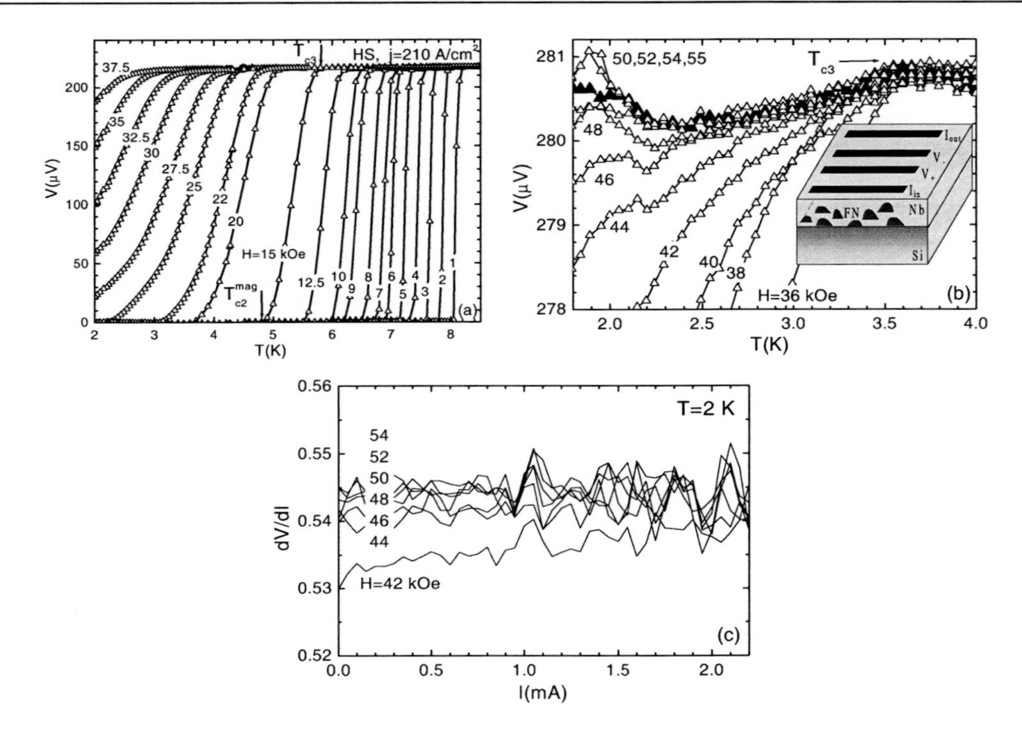

Figure 6: (a) Detailed voltage curves in the HS as a function of temperature for various magnetic fields under a dc current density $j_{\rm dc} = 210$ A/cm^2. (b) Representative curves near the normal state and for high applied fields $H > 36$ kOe where superconductivity is gradually weakened. We observe that the T_{c3} point is almost constant, the curves are broadened and gradually a local minimum is formed at $T \approx 2.4$ K. The filled triangles represent the data for $H = 55$ kOe (maximum accessible field). The HS and the configuration of the transport measurements are presented schematically in the inset. The contact pads are made on the top surface of the Nb layer. (c) We focus on the dV/dI vs I curves obtained at $T = 2$ K and in the whole high-field range accessible in our SQUID. In all cases the magnetic field was parallel to the easy-axis $\hat{a}_e$ of magnetization ($\mathbf{H} \parallel \hat{a}_e$).

point $T_{c3}(H)$ is exceeded. Furthermore, the same behaviour has been observed in measurements of the $I - V$ characteristics performed in the regime under discussion. Such data are presented in figure 7 (c) in a logarithmic plot for applied field $H = 25$ kOe at various temperatures. In this plot the dotted lines represent the normal state curve (referring to $T = 5$ K$> T_{c3}(25$ kOe$)$). These results reveal straightforwardly the non-linear behaviour in the regime above $T_{c2}^{\rm mag}(H)$. The inset presents the respective $V(T)$ curve for $H = 25$ kOe. By taking into account the comparison of the resistance data with the magnetization data and in agreement to recently published experimental works [37, 38, 23, 26] we attribute the regime $T_{c2}^{\rm mag}(H) < T < T_{c3}(H)$ to surface superconductivity (see also below) [39, 40]. It should be noted that a clear structure is observed in the resistance curves when very low transport currents are applied (see figure 7(a) for $I_{\rm dc} = 10$ and 25 μA). This structure is observed in the regime of the bulk upper-critical field $H_{c2}(T)$. In agreement to recent works where hybrid films of ordered arrays of magnetic particles embedded in a superconductor have been studied we may attribute this effect to the matching of the vortex lattice which is hosted in the Nb layer with the number of the randomly distributed magnetic particles.

The characteristic points for the nucleation of superconductivity in our HS are summarized in figures 8(a) and 8(b). Figure 8(a) presents the surface superconductivity fields $H_{c3}(T)$ (filled circles) and the bulk upper-critical fields $H_{c2}(T)$ as determined from magnetic (open squares) and resistive (open circles) measurements. In addition, the $H_{c3}(T)$ line of the pure Nb film is represented by the thick solid line. In figure 8(b) we present in more detail the magnetic data corresponding to the bulk upper-critical field line $H_{c2}(T)$. The dotted lines refer to extrapolations of the middle-field $H_{c2}(T)$ data to zero temperature (main panel) and zero field (inset). These data will be discussed analytically below in comparison to recent theoretical proposals. It should be noted that the data presented in figures 8(a) and 8(b) were obtained by performing successive measurements as a function of temperature at constant external magnetic field according to the following experimental protocol: initially the HS was carefully demagnetized. In every new temperature scan the applied magnetic field was always higher than the value that it had during the previous measurement. In case where a measurement had to be performed at a magnetic field value lower than the last one the HS was demagnetized prior to the next measurement. This procedure is needed because the magnetic *memory effects* that are observed in our HS could significantly complicate the behaviour of the $H_{c2}(T)$ and $H_{c3}(T)$ boundary lines presented in figures 8(a) and 8(b) (see below). From these data a number of interesting features may be noticed. First, while the $H_{c3}(T)$ lines of the HS and reference Nb film coincide in the low-field regime, at a characteristic field $H \approx 8$ kOe the data of the HS exhibit a clear change in the slope and the two lines start to diverge. This characteristic field is equal to the saturation field $H_{sat}^{soft} \simeq 8$ kOe of the SMNs (see figure 4 (a)). Since the HMNs exhibit a significantly higher saturation field $H_{sat}^{hard} \simeq 36$ kOe we expect that an analogous effect should also be observed in the high-field regime. Indeed, the high-field $H_{c3}(T)$ data present a pronounced upturn at $H_{sat}^{hard} \simeq 36$ kOe. By recalling the data of figure 6(b) we note that magnetic fields $H > H_{sat}^{hard} \simeq 36$ kOe don't shift the point T_{c3} where the normal state resistance is attained but only broaden the voltage curves which gradually tend to become horizontal. Returning to figure 8 (a), the boundary where the HS reenters the normal state is also introduced (solid squares) in the low-temperature regime (shaded area). It should be noted that the reentrance observed in the HS is not observed in the reference pure Nb film which was prepared simultaneously during the same deposition process. Despite that we are not able to perform measurements in the regime $T < 1.8$ K, $H > 55$ kOe we believe that the reentrance observed in our HS is related to the normal metallic state and not to a field-driven superconductor-insulator transition. This is indicated by the $I - V$ data presented in figure 6(c). This may be further supported if we take into account that, in the past, a field-driven superconductor-insulator transition has been observed only in samples of very low thickness d (typically $d < 10$ nm) which exhibit high normal state resistance [41, 36]. The Nb layer in our HS is thick ($d = 200$ nm) and as a consequence exhibits three dimensional behaviour. In addition, in two dimensional Nb films [41, 36] the magnetoresistance curves at the superconductor-insulator transition extend in a wide temperature regime since their increase is initiated fairly above the zero-field critical temperature of the superconductor. In contrast, in our HS the normal state is recovered only below the characteristic temperature $T \approx 2.4$ K and is *bounded by the surface superconducting regime*.

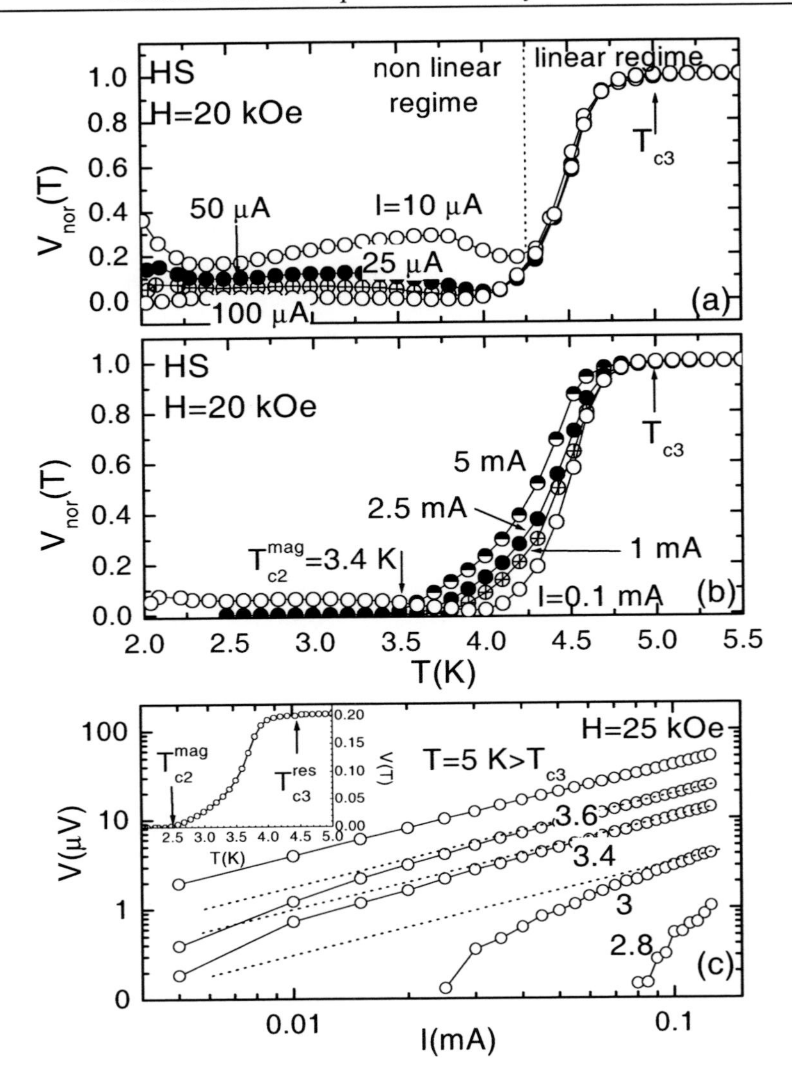

Figure 7: Representative normalized resistance curves for low $I_{\rm dc} = 10, 25, 50$ and 100 μA (upper panel) and high $I_{\rm dc} = 0.1, 1, 2.5$ and 5 mA (middle panel) transport currents at $H = 20$ kOe. While for low currents the regime close to the normal state is linear for higher current densities the non-linear behaviour of the HS is revealed in the regime $T_{\rm c2}^{\rm mag}(H) < T < T_{\rm c3}(H)$. In the lower panel the $I - V$ characteristics are presented where a clear non-linear behaviour is observed in the regime above the magnetically determined bulk upper-critical points $T_{\rm c2}^{\rm mag}$. The dotted lines refer to the slope of the linear normal state voltage curve obtained at $T = 5$ K> $T_{c3}(20$ kOe$)$ The $I - V$ curves attain the linear character only when the characteristic point $T_{\rm c3}$ is exceeded. The inset shows the respective $V(T)$ curve for $H = 20$ kOe. In all cases the magnetic field was parallel to the easy-axis $\hat{a}_e$ of magnetization ($\mathbf{H} \parallel \hat{a}_e$).

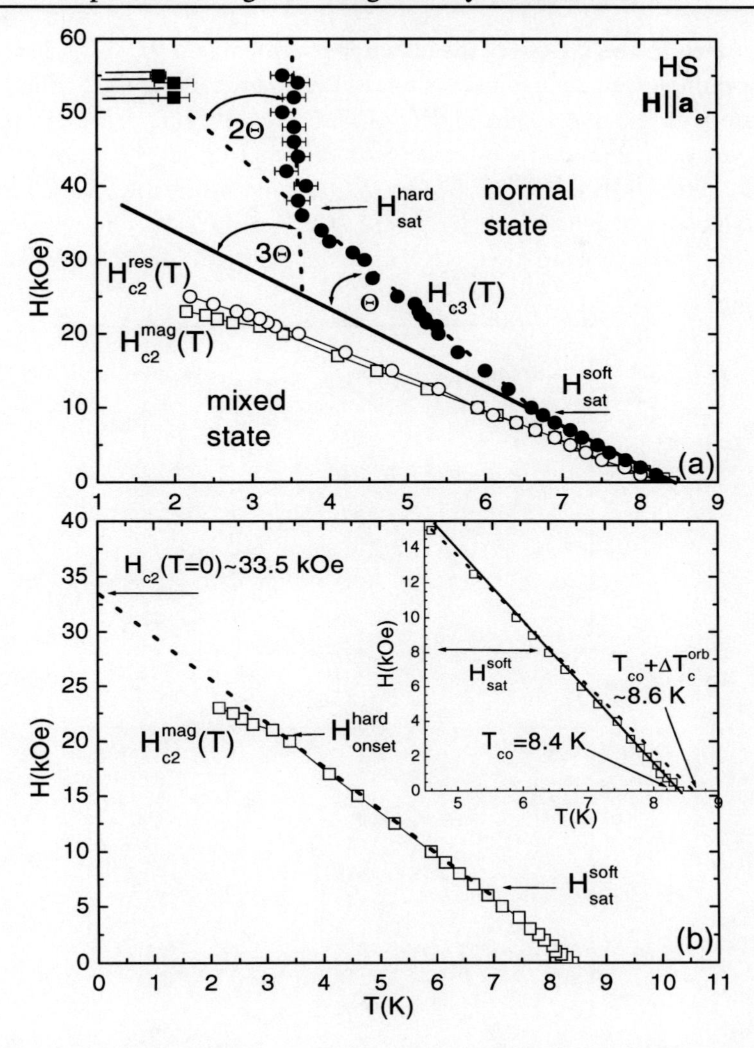

Figure 8: (Upper panel) Magnetic-field-temperature boundaries indicating bulk (open circles and squares) and surface (filled circles) nucleation of superconductivity as constructed from magnetic and resistance measurements. In addition, the superconducting-normal boundary is presented at $T \approx 2$ K (filled squares). Clear changes in the slope of $H_{c3}(T)$ and $H_{c2}(T)$ are observed at some characteristic field values. 3Θ and Θ indicate the angles formed by the high-field and middle-field part of the $H_{c3}(T)$ line with its low-field extrapolation. The $H_{c3}(T)$ points of the Nb reference film follow the thick solid line [23]. (Lower panel) Magnetically determined bulk upper-critical fields $H_{c2}(T)$. The inset focuses in the low-field regime where a weak change in the slope of $H_{c2}(T)$ is observed. Dotted lines refer to the middle-field part of $H_{c2}(T)$ when extrapolated to zero temperature (main panel) and zero field (inset). For more details see the discussion in the text. The results refer to the case where the magnetic field was applied parallel to the easy-axis $\hat{a}_e$ of magnetization ($\mathbf{H} \parallel \hat{a}_e$).

We now discuss the behaviour of the bulk upper-critical line $H_{c2}(T)$. This characteristic line exhibits a pronounced change in its slope at the characteristic field value $H \approx 20$ kOe, which correlate with the onset field $H_{\mathrm{onset}}^{\mathrm{hard}}$ of HMNs (see figure 4(a)). As may be seen in the inset of figure 8(b), a second weak change in the slope of $H_{c2}(T)$ line is observed at the saturation field of the SMNs $H_{\mathrm{sat}}^{\mathrm{soft}} \approx 8$ kOe. We note that while the $H_{c3}(T)$ line exhibits an *increase* at both characteristic fields $H_{\mathrm{sat}}^{\mathrm{soft}}$ and $H_{\mathrm{sat}}^{\mathrm{hard}}$, the $H_{c2}(T)$ line presents a *decrease* in its slope at the two characteristic fields $H_{\mathrm{sat}}^{\mathrm{soft}}$ and $H_{\mathrm{onset}}^{\mathrm{hard}}$.

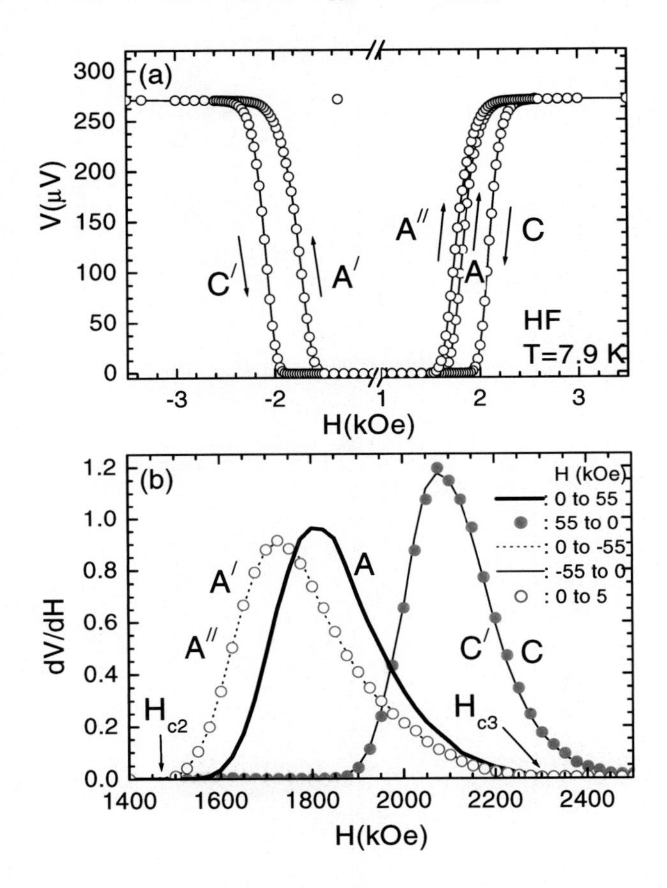

Figure 9: Voltage curves at $T = 7.9$ K as a function of the magnetic field. (Upper panel) We focus on the regime of the transition but notice that the curves were recorded in the whole field range -55 kOe$< H < 55$ kOe. (Lower panel) Comparative presentation of all respective derivatives $dV(H)/dH$ of the obtained curves for positive field values. The two curves $A^/$ and $C^/$ (originally measured at negative fields) are reversed for the sake of comparison. We clearly observe that the curves $A^/ - A^{//}$ and $C - C^/$ coincide completely in the entire field range. In addition, the bulk upper-critical field H_{c2} and the surface superconductivity field H_{c3} strongly depend on the magnetic history of the HS. The applied magnetic field was parallel to the easy-axis $\hat{a}_e$ of magnetization ($\mathbf{H} \parallel \hat{a}_e$).

The change in the slope observed in the $H_{c2}(T)$ line at low fields is very weak and probably could be questioned. In order to leave no doubt, we performed some specific magnetic and magnetoresistance measurements which were especially designed in order to prove that

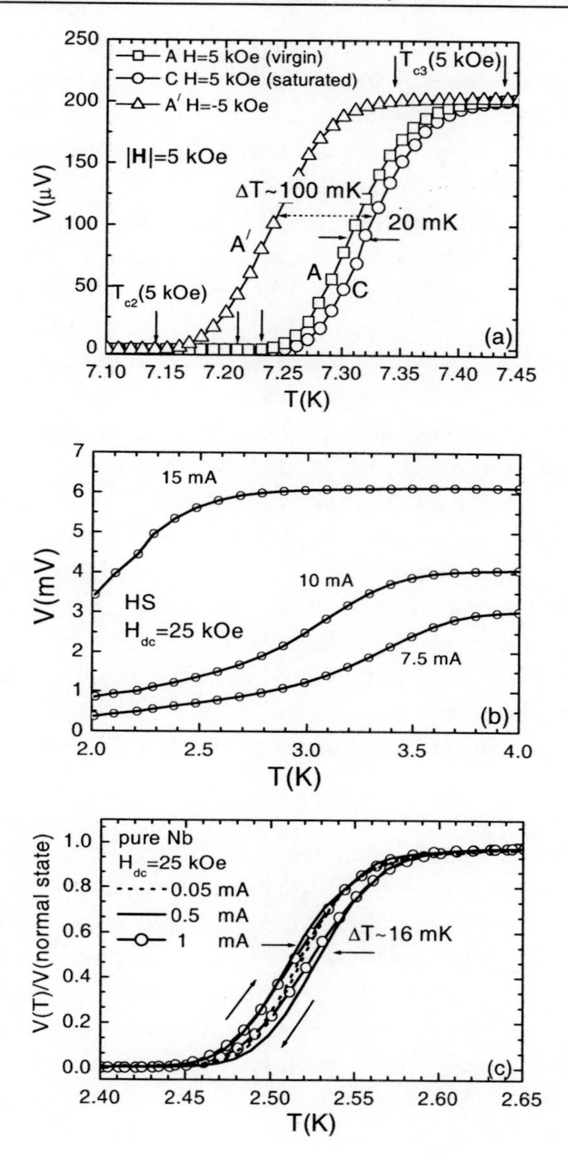

Figure 10: (a) Isofield voltage curves obtained as a function of temperature at $H_{dc} = 5$ kOe. The temperature shift of the obtained curves is of the order $\Delta T_c \simeq 100$ mK. (b) Sequential isofield voltage curves (open points and solid lines) obtained for a HS as a function of temperature at $H_{dc} = 25$ kOe for various transport currents. Notice that in this temperature range the obtained temperature shift between the respective curves is not visible since it is only a few mK. (c) Isofield voltage curves obtained in a pure Nb film for sweeping the temperature at $H_{dc} = 25$ kOe. The maximum temperature shift of the obtained curves owing to the conventional undercooling and superheating at the transition depends on the applied current and is well below 20 mK. For low magnetic fields the temperature shift is in the range of mK. For the data obtained in the HSs the applied magnetic field was parallel to the easy-axis $\hat{a}_e$ of magnetization ($\mathbf{H} \parallel \hat{a}_e$).

the observed effect is real and is actually motivated by the presence of the MNs. Representative data are shown in figures 9(a) and (b) for $T = 7.9$ K. The experimental protocol of the performed measurements is as following: Initially the HS is carefully demagnetized at $T = 10$ K $> T_c = 8.4$ K by performing sequential loops while decreasing the maximum value of the applied magnetic field. After this procedure is completed the magnetic moment of the MNs exhibits a typical value $m < 5 \cdot 10^{-6}$ emu. We then lower the temperature at $T = 7.9$ K and we measure the magnetoresistance as a function of the applied field from 0 kOe$\longrightarrow$ 55 kOe (curve A), 55 kOe$\longrightarrow$ 0 kOe (curve C), 0 kOe$\longrightarrow$ -55 kOe (curve $A^{/}$), -55 kOe$\longrightarrow$ 0 kOe (curve $C^{/}$) and finally 0 kOe$\longrightarrow$ 5 kOe (curve $A^{//}$). At the end of this procedure an entire loop of the MNs has been performed. In figure 9 (a) we present the raw data in the low-field regime close to the transition. Note that the voltage curves were recorded in the whole range of magnetic fields accessible in our SQUID -55 kOe$< H < 55$ kOe. In figure 9 (b) we present the derivative $dV(H)/dH$ for all curves. We have to stress that the curves $A^{/}$ (dotted lines) and $C^{/}$ (thin solid line) are reversed to positive field values for the sake of comparison with the curves A (thick solid line), C (solid circles) and $A^{//}$ (open circles). We clearly see that the magnetic behaviour of the MNs is reflected in the superconducting properties of the Nb layer. This is evident since both the bulk upper-critical field H_{c2} and the surface superconductivity field H_{c3} depend on the magnetic history. The maximum relative shift between the obtained curves is of the order $\Delta H = 400$ Oe. The curves $C - C^{/}$ and $A^{/} - A^{//}$ coincide absolutely in the entire field regime. We will refer extensively to these results in a following section of the present chapter.

It should be noted that the same behaviour is observed in isofield voltage data $V(T)$ as a function of temperature. Representative data are shown in figure 10(a) for $|H_{dc}| = 5$ kOe when the HS was initially virgin (squares), saturated (circles) and when the field was reversed (triangles). The observed temperature shift between the obtained curves is $\Delta T_c \simeq 100$ mK. In order to stress that the observed memory effect is directly related to the existence of MNs and is not motivated by a limited temperature resolution or stabilization inability of our SQUID in figure 10(b) we present successive measurements performed exactly under the same experimental conditions. The respective curves almost coincide exhibiting a temperature shift in the range of few mK. In the same figure 10(c) we also present data for sweeping the temperature for a pure Nb film in order to point out that the effect under discussion is not related to the conventional undercooling and superheating which is usually observed at the superconducting-normal transition. For pure Nb films the maximum temperature hysteresis was observed for an applied magnetic field $H_{dc} = 20$ kOe and is limited well below 20 mK, while is entirely absent in the low-field regime. Thus, we safely conclude that the pronounced memory effects observed in the HS are motivated by the MNs.

3.2 Second Category of HSs: MNs Preferably Embedded in Half of a Nb Layer

In the second category of samples presented in this chapter the MNs are preferably embedded in half of a high-quality Nb layer as is schematically presented at the inset of figure 11. Thus the whole film is a combinatorial structure (CS) which consists of a pure Nb layer adjacent to a HS. In figure 11 are shown representative voltage curves, measured at the two different sites of the CS, as a function of temperature for various magnetic fields applied parallel to the easy-axis $\hat{a}_e$ of magnetization ($\mathbf{H} \parallel \hat{a}_e$). The curves refer to the voltage $V_{1,4}(T)$ which was measured between the characteristic points 1 and 4. As is presented schematically in the inset these points are positioned on different parts of the CS. Point 1 is placed on the superconducting site, while point 4 is positioned on the HS part of the CS. Since the MNs are initially demagnetized, in zero magnetic field the voltage curve should not exhibit any special feature as indeed is evident in the data. We note that by employing very long pre-sputtering times prior to the actual deposition we diminished the oxygen residue in the chamber so that the deposited Nb layer is of high quality and maintains homogenous superconducting properties throughout the CS [25]. This is evident by its high zero-field $T_c = 8.41$ K and the sharpness of its single-step transition which exhibits $\Delta T \approx 50$ mK according to a $10\% - 90\%$ criterion (see also figure 14 below). In contrast to zero-field data, when an external magnetic field is applied the situation should alter dramatically due to the presence of the MNs in the HS part. As the external field is increased any modulation in the critical temperature of the HS in respect to the adjacent Nb layer should result in a structure in the measured curves $V_{1,4}(T)$ indicative of two different transitions. Indeed, this behavior is clearly observed in the data presented in figure 11. As we perform the measurements in higher external fields the voltage curves $V_{1,4}(T)$ broaden and gradually present a two step feature which indicates the two different transitions of the two different parts of the CS. We observe that as the temperature decreases the $V_{1,4}(T)$ curves deviate from the normal state value at points $T_{c3}(H)$. In addition the $V_{1,4}(T)$ curves exhibit first a change in their slope at a field-independent voltage level $V_{1,4}(T) \simeq 80\mu V$ (see horizontal dotted line tracing the points $T_{cs}(H)$) and second a sharp drop towards zero at a field dependent voltage level (see inclined dashed line tracing the points $T_{sd}(H)$). Finally, the CS becomes totaly superconducting at points $T_{c2}(H)$ where the resistance gets zero.

The direct comparison of $V_{1,4}(T)$ with $V_{3,4}(T)$ and $V_{1,2}(T)$ revealed that the upper part of the voltage curves $V_{1,4}(T)$ refers to the transition of the HS, while their lower part refers to the transition of the adjacent Nb layer. Representative data are shown in figure 12(a) where we present comparatively the voltage curves $V_{1,4}(T)$ and $V_{3,4}(T)$ for magnetic fields $H = 10, 16$ and 22 kOe applied parallel to the easy-axis $\hat{a}_e$ of magnetization of the MNs. First of all we see that as the temperature decreases both voltage curves $V_{1,4}(T)$ and $V_{3,4}(T)$ deviate from the normal state value at the same characteristic points $T_{c3}(H)$. It should be noted that in figure 12(a) it appears that, for $H = 10$ and 16 kOe, $V_{3,4}(T) > V_{1,4}(T)$ near the onset point, while from the schematic inset of figure 11 it is expected that $V_{3,4}(T) < V_{1,4}(T)$ should always hold. This is not actual, but factitious. During our measurements only two curves, namely the $V_{1,2}(T)$ and $V_{3,4}(T)$, could be measured simultaneously. The measurements of the curves $V_{1,4}(T)$ were performed in different scans. Thus, the observed irregularity is related to a small relative temperature shift between $V_{3,4}(T)$ and $V_{1,4}(T)$ due to the slightly different values of the applied magnetic field during different measurements (no-overshoot operation mode of the SQUID). The data presented in figure 12(b), that were obtained when $V_{1,2}(T)$ and $V_{3,4}(T)$ were measured *simultaneously*, leave no doubt for the validity of our results. Returning to figure 12 (a) we observe that the magnetically determined bulk upper-critical temperatures $T_{c2}^{\mathrm{HS}}(H)$ of the HS part, represented by the zeroing of the $V_{3,4}(T)$ curves, nicely correlate with the points where the

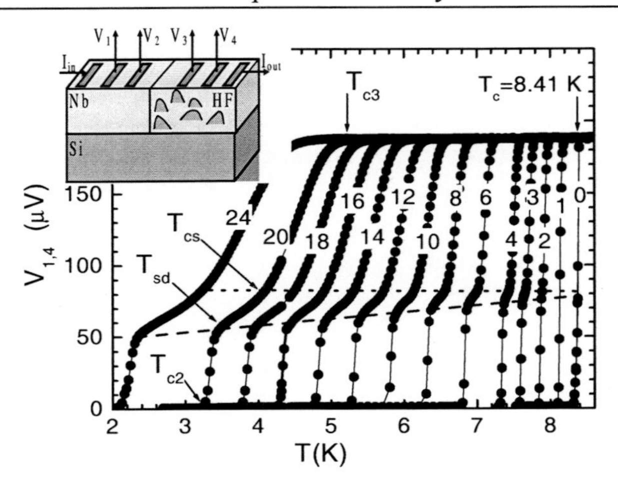

Figure 11: Detailed measurements of the voltage $V_{1,4}(T)$ for various magnetic fields 0 kOe$\leq H \leq$ 24 kOe. The inset presents schematically the CS and the configuration of the performed measurements. In all cases the applied field was parallel to the magnetization easy-axis $\hat{a}_e$ of MNs.

sharp drop in the voltage curves $V_{1,4}(T)$ initiates. We are now in a position to identify all four characteristic points existing in the $V_{1,4}(T)$ curves: The points $T_{\rm c3}(H)$ where the normal state value is obtained refer to the points where surface-like superconductivity starts to form in the HS part of the CS. The points $T_{\rm sd}(H)$ where the sharp drop toward zero initiates (see inclined dashed line) are related to the bulk upper-critical temperatures $T_{\rm c2}^{\rm HS}(H)$ of the HS part. The zeroing temperatures reflect the points $T_{\rm c2}^{\rm Nb}(H)$ where bulk superconductivity is established in the Nb part of the CS (then superconductivity is maintained in the whole CS) and finally the points $T_{\rm cs}(H)$ where a change in the slope is observed in the $V_{1,4}(T)$ curves (see horizontal dotted line) refer to the temperatures $T_{\rm c3}^{\rm Nb}(H)$ where the pure Nb region of the CS enters the normal state as we increase the temperature [23, 24, 25, 26]. *An important outcome of the data presented so far is that in the HS the superconducting transition takes place at higher temperatures than in the pure Nb area.* In the lower panel, figure 12(b) we focus near the normal state boundary of the $V_{1,2}(T)$ (open triangles) and $V_{3,4}(T)$ (solid triangles) voltage curves which were measured at the two different parts of a second CS. These data show clearly that, for the same applied magnetic field and current, the normal state boundary of the Nb part is placed at lower temperatures when compared to the HS part of the CS. Since all external experimental parameters (temperature, external magnetic field and applied current) are exactly the same for the Nb layer existing throughout the CS the enhanced superconducting temperature of its HS part should be attributed exclusively to the influence of the MNs [23, 24, 25, 26].

The characteristic points discussed above are summarized in figure 13 for the field configuration $\mathbf{H} \parallel \hat{a}_e$. We observe that the line $H_{\rm c3}^{\rm HS}(T)$, which designates the normal state of the HS part of the CS, exhibits a change in its slope at the saturation field $H_{\rm sat}^{\rm MN} \simeq 5$ kOe of the MNs (see the inset where the initial part of the virgin magnetization loop is presented for $T > T_c$). All other boundary lines referring to the different characteristic points don't exhibit such a tendency, but clearly maintain an almost linear behavior in the

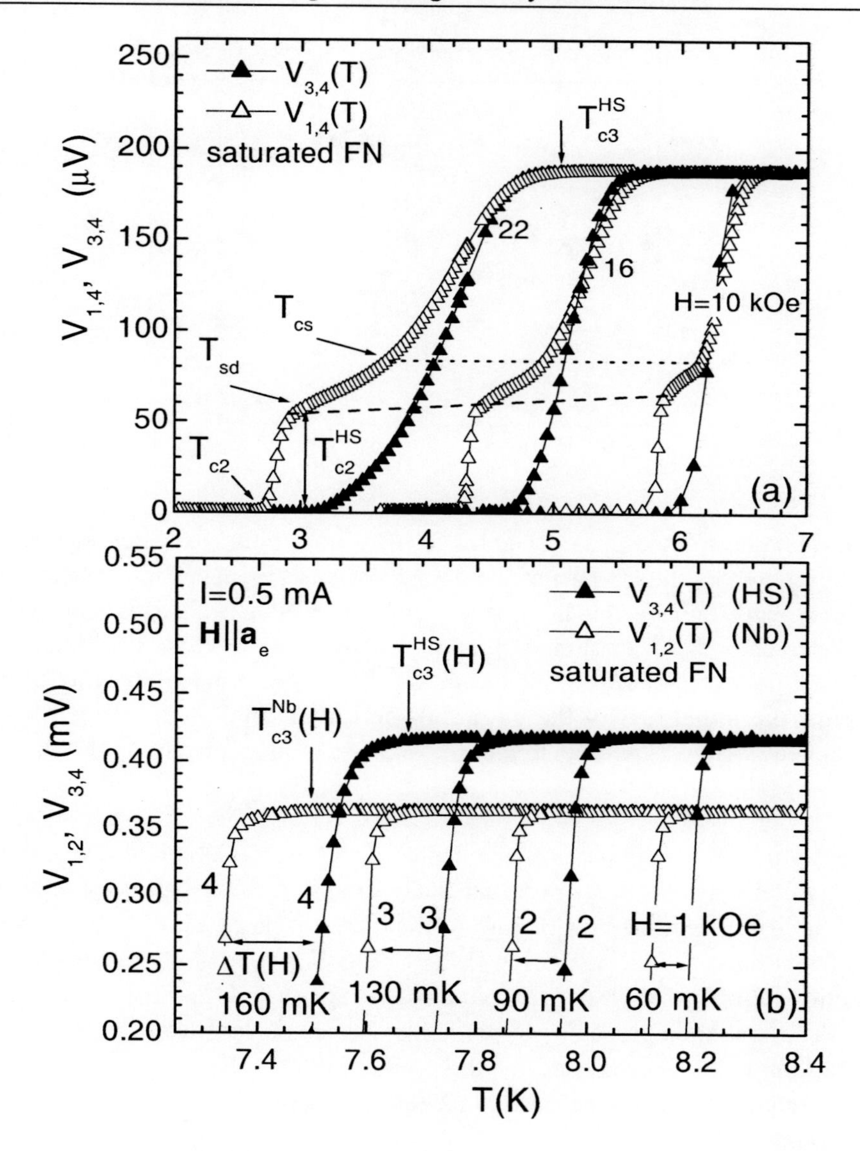

Figure 12: (Upper panel) Comparative presentation of voltage curves $V_{1,4}(T)$ (open triangles) and $V_{3,4}(T)$ (solid triangles) at external magnetic fields $H = 10, 16$ and 22 kOe. (Lower panel) Comparative presentation of voltage curves $V_{1,2}(T)$ (open triangles) and $V_{3,4}(T)$ (solid triangles) measured in a second CS at low magnetic fields $H = 1, 2, 3$ and 4 kOe. We observe that as the magnetic field increases the deviation of the respective voltage curves $V_{1,2}(T)$ and $V_{3,4}(T)$ is getting more pronounced. In all cases the applied field was parallel to the magnetization easy-axis $\hat{a}_e$ of MNs.

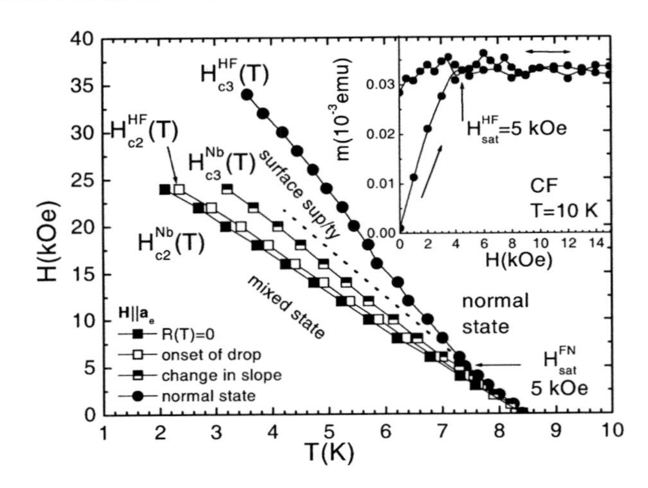

Figure 13: The constructed phase diagram for the nucleation of superconductivity in our CS. The regime where the HS part of the CS maintains a resistance lower than the normal state value is greatly enhanced in comparison to the pure Nb part. This is achieved since an upturn is observed in $H_{c3}^{\rm HS}(T)$ at the saturation field $H_{\rm sat}^{\rm MN} \simeq 5$ kOe of the MNs. The dotted line represents the extrapolation of the low-field $H_{c3}^{\rm HS}(T)$ data in high magnetic fields. The inset presents the initial part of the virgin magnetization loop at $T = 10$ K$> T_c$. The applied magnetic field was parallel to the easy-axis $\hat{a}_e$ of magnetization ($\mathbf{H} \parallel \hat{a}_e$).

whole regime investigated in the present chapter. As a result the regime where the HS part of the CS maintains a resistance lower than the normal state value is significantly enhanced when compared to the respective regime of the pure Nb part. This finding suggests that by controlled variation of the saturation field of the MNs we may directly enhance the superconducting regime of the $H - T$ operational diagram [24, 25, 26]. It should also be noted that the preparation procedure of randomly-distributed isolated MNs is much simpler compared to the techniques needed for the production of ordered ones. Thus, randomly distributed MNs could be more attractive for the construction of HSs that will be used for current-carrying applications.

The presence of MNs indeed motivate magnetic memory effects in the superconducting state of the HS as demonstrated by the data presented above for the case of simple HSs. To investigate the existence of such phenomena in the second category of CSs presented in this chapter we performed measurements based on different magnetic histories of the MNs. Such data are presented in figure 14. In this figure solid points refer to the curves $V_{1,4}(T)$ obtained when the MNs were initially carefully demagnetized, while the open points (positive fields) and points with crosses (negative fields) refer to $V_{1,4}(T)$ data obtained when initially the MNs were saturated by applying a magnetic field $H > H_{\rm sat}^{\rm MN} \simeq 5$ kOe. We clearly see that in the temperature regime $T_{c2}^{\rm HS}(H) < T < T_{c3}^{\rm HS}(H)$ the voltage curves $V_{1,4}(T)$ are greatly affected by the magnetic state of the MNs, while their lower part $T_{c2}^{\rm Nb}(H) < T < T_{c2}^{\rm HS}(H)$ is left unchanged. This is so because the lower segment $T_{c2}^{\rm Nb}(H) < T < T_{c2}^{\rm HS}(H)$ of the curves is related to the transition of the pure Nb layer of the CS, while the upper segment $T_{c2}^{\rm HS}(H) < T < T_{c3}^{\rm HS}(H)$ reflects the transition of the

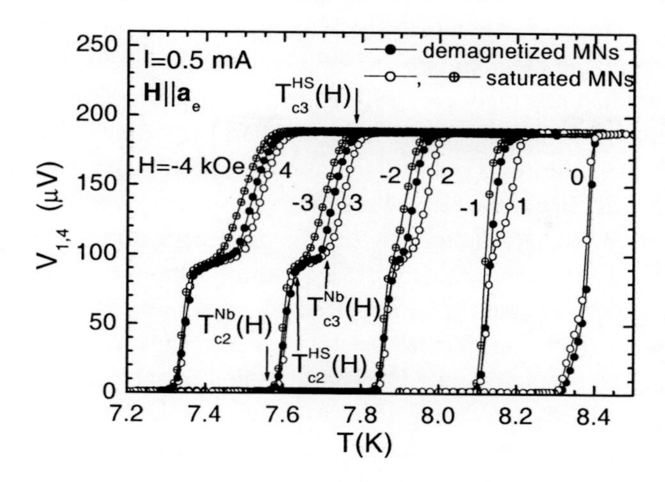

Figure 14: Voltage curves $V_{1,4}(T)$ at magnetic fields $H = 0, \pm 1, \pm 2, \pm 3$ and ± 4 kOe after demagnetization (solid points) and after saturation (open points and points with crosses) of the MNs. Magnetic memory effects are observed in the $V_{1,4}(T)$ curves only in the temperature regime $T_{c2}^{\mathrm{HS}}(H) < T < T_{c3}^{\mathrm{HS}}(H)$, while their lower parts $T_{c2}^{\mathrm{Nb}}(H) < T < T_{c2}^{\mathrm{HS}}(H)$ coincide entirely. In all cases the magnetic field is parallel to the magnetization easy-axis $\hat{a}_e$ of the MNs ($\mathbf{H} \parallel \hat{a}_e$).

HS part of the CS. *Since the superconducting Nb layer extending in the whole CS is subjected to exactly the same extrinsic parameters (temperature, external magnetic field and applied current) it is only the MNs that could motivate the modulation of superconductivity and the magnetic memory effects observed in its HS part.* We observe that the curves obtained at positive fields after the saturation of the MNs are placed in higher temperatures compared to the curves obtained when the MNs were initially demagnetized or the ones obtained at negative field values. This result clearly proves that the nucleation temperature of superconductivity can be modulated in a controlled way by the action of MNs.

4 Comparison with Theoretical Proposals and Interpretation of the Experimental Results

4.1 Theoretical Overview

The results presented in the previous section may be explained by simple considerations as discussed below after a brief overview of the relevant theoretical background. In hybrid systems (in our case having the form of films) there are two basic mechanisms that control the interaction between the superconducting order parameter and the magnetic moments. First, is the electromagnetic mechanism which is related to the interaction of the superconducting pairs with the magnetic fields induced by the ferromagnet in the superconductor and second is the exchange interaction that the superconducting pairs experience as they enter the ferromagnetic areas [42, 43, 44, 45, 34]. Under specific circumstances both mechanisms may promote or act against superconducting order. Regarding the electromagnetic mechanism, there are cases where the internal fields produced by the ferromagnetic ingredient compensate the external applied field at some regions of the HS where the two components are directed reversely [35, 19, 23, 24, 25, 26]. Thus, at some areas of the HS where the total effective magnetic field is lower than the external field, superconductivity will survive at temperatures higher than that observed in a single superconducting film [35, 19, 23, 24, 25, 26]. In contrast, there are also areas in the superconducting layer of the HS where the internal fields (produced by the ferromagnetic component) are added to the external field. As a result, these areas of the HS exhibit suppressed superconducting properties. On the other hand, the pair-breaking effect, due to the exchange field that the Cooper pairs experience when they enter ferromagnetic areas, strongly suppresses superconducting order [45, 46, 47, 34]. It was first noted by Matthias and Suhl [48] that enhancement of the superconducting properties should only occur in the vicinity of domain walls than inside the domains of a ferromagnet. In recent years this mechanism of superconductivity formation in ferromagnetic superconductors or HSs was studied extensively. A. I. Buzdin and coworkers showed that this mechanism is analogous to the one responsible for the formation of surface superconductivity at the interface of a superconductor with an insulator [44, 45, 34]. Very recently Rusanov et al. showed that the pair-breaking effect due to the exchange field may be suppressed when the Cooper pairs sample different directions of the exchange field [49]. Since this condition favorably occurs at the domain walls of a ferromagnet the effect was called enhancement of T_c induced by the domain walls [49]. In our case the second mechanism which attributes the enhancement of superconductivity to the presence of domain walls is not favored to occur or have little contribution to the effect observed in our data. Since the MNs have small dimensions it is more probable that each one is a single magnetic domain, thus domain walls don't exist. In addition, even in the case where the MNs have a multi-domain magnetic structure the observed effect could not be ascribed to the presence of domain walls since it clearly occurs in the regime above each saturation field $H_{\rm sat}^{\rm soft}$ or $H_{\rm sat}^{\rm hard}$ where domain walls are no longer present. Finally, due to their specific structure in our HSs there are large areas of the superconducting layer that are exposed to the dipolar fields of the MNs. This condition promotes the action of the electromagnetic mechanism. Here we have to underline that due to the high demagnetization factor of the MNs their dipolar fields are mainly restricted in the regimes near their lateral surfaces at a distance of the order of their thickness ($d \simeq 10 - 50$ nm). Thus, the compensation of the external magnetic field is efficient mainly in these regimes. As a result the superconducting order parameter, promoted under the action of the internal dipolar fields, should be also restricted very close to the lateral surfaces of the MNs. *Since this spatially modulated superconducting order parameter is reminiscent of surface superconductivity (obtained at a conventional superconductor-insulator interface) we may generalize and also call the effect under discussion surface-like superconductivity.* This notation is clearly further supported by the experimental results presented in the previous sections of this chapter.

4.2 Memory Effects

Due to the reasons outlined above we believe that in our case the nucleation of superconductivity is rather controlled by the action of the electromagnetic mechanism than by the presence of magnetic domain walls. Under this assumption let us now proceed with the interpretation of our experimental results. Initially we will refer to the first category of HSs. The respective data of magnetic memory effects are presented in figures 9(a)-(b) in the form of raw $V(H)$ curves and their derivatives $dV(H)/dH$. The situation needed for the occurrence of memory effects is illustrated in figure 15. Initially, the MNs are demagnetized (randomly oriented) at $T = 10$ K> $T_c = 8.4$ K and their contribution on the superconductor is negligible (panel (a) of figure 15). We set the temperature at $T = 7.9$ K and start increase the external field from $0 \longrightarrow 55$ kOe. During this measurement we detect curve A (see figures 9 (a)-(b)). As the applied field increases the MNs are gradually oriented (panel (b) of figure 15). Their internal dipolar fields $\mathbf{H}_{in}$ enhance the external field $\mathbf{H}_{ex}$ in the regimes above their normal surfaces, while reduce it in the regimes beside their lateral surfaces [35, 19, 23, 24, 25, 26]. As a consequence in the regime by the lateral surfaces of the MNs superconductivity is preserved at temperatures higher than should be expected in the absence of MNs. Thus, the $T_{c2}(H)$ and $T_{c3}(H)$ should be slightly increased comparing to pure Nb. Although the $T_{c3}(H)$ is exceeded we keep increase the external magnetic field above the saturation field $H_{\rm sat}^{\rm hard} \simeq 36$ kOe of the HMNs and subsequently decrease it. The respective data measured for $55 \longrightarrow 0$ kOe are represented by curve C (see figures 9 (a)-(b)). After this procedure is completed all the MNs are in the remanent state (panel (c) of figure 15). Now the external magnetic field $\mathbf{H}_{ex}$ is suppressed more efficiently in the whole film's area due to the increased negative contribution by the dipolar internal fields $\mathbf{H}_{in}$ of all MNs. As a result superconductivity should be preserved at even higher fields. This is actually observed in figures 9 (a)-(b) where clearly curve C is placed at higher magnetic fields compared to curve A.

In contrast, when we reverse the external field (curve $A^/$ in figures 9 (a)-(b)) the MNs also reverse their internal dipolar fields. Nevertheless, until the coercive field $H_c = 13$ kOe is exceeded most of them preserve a positive magnetization as may be seen directly in the magnetization data of figure 4. The situation is schematically presented in panel (d) of figure 15. As a consequence their internal dipolar fields $\mathbf{H}_{in}$ now enhance the external magnetic field $\mathbf{H}_{ex}$ (rather than reduce it as discussed above) in the regimes beside their lateral surfaces. As a consequence now superconductivity should be destroyed at a lower critical field comparing to the previous measurements. This behaviour is clearly exhibited by the voltage curves in figure 9 (b) where the derivative $dV(H)/dH$ of curve $A^/$ is reversed to positive fields for the sake of comparison. Since in all three cases (curves A, C and $A^/$) the temperature was constant at $T = 7.9$ K the different critical fields $H_{c2}(7.9\mathrm{K})$ and $H_{c3}(7.9\mathrm{K})$ that were observed could only be attributed to the different internal fields that are produced by the MNs. Finally, due to simple considerations based on the symmetry exhibited in the magnetic loop (see figure 4(a)), one should expect that the behaviour observed in curves $A^{//}$ and $C^/$ should be exactly the same with the one observed by $A^/$ and C respectively. Indeed, this is confirmed by the entire concurrence of the respective derivative curves $dV(H)/dH$ presented in figure 9 (b).

By careful inspection of panel (d) of figure 15 one could object that enhancement of superconductivity should also be observed when the external magnetic field is reversed (see curve $A^/$ in figures 9 (a)-(b)) since the external magnetic field $\mathbf{H}_{ex}$ is reduced, not in the regimes *beside* the lateral surfaces of MNs as assumed for the case of panels (b)-(c) of figure 15, but in the regimes *above* the MNs. The question that naturally arises is why don't we observe enhancement of superconductivity in this case? The answer we want to offer

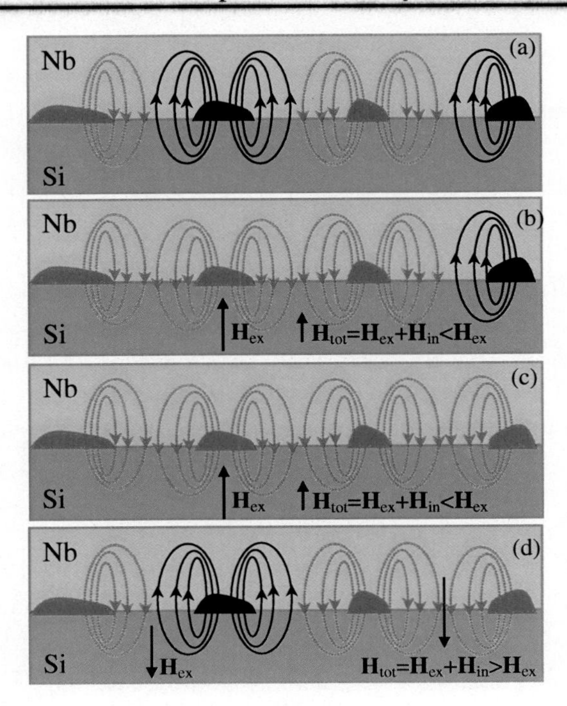

Figure 15: Schematic representation of the experimental conditions for the occurrence of memory effects. Panel (a) refers to the demagnetized state of the MNs, while panels (b), (c) and (d) refer to the situations where curves A, C and $A^{/}$ of figure 9 have been obtained. We observe that in the regimes by the lateral surfaces of MNs the external magnetic field $\mathbf{H}_{ex}$ may be reduced or enhanced at will by their internal dipolar fields $\mathbf{H}_{in}$.

is the following and is based on the conditions that are schematically presented in figures 16(a)-(b) where the side and top views of the HS may be seen. More specifically in the top view of the HS presented in the lower panel of figure 16 we observe that the regimes where the external field $\mathbf{H}_{ex}$ is effectively compensated by the MNs' internal fields $\mathbf{H}_{in}$ are placed right above their top surfaces. Notice that these regimes where superconductivity should now be enhanced are **isolated** from each other since they are surrounded by regimes where the total field $\mathbf{H}_{tot}$ is higher that the external one $\mathbf{H}_{ex}$. As a result in this case a transport current could not avoid the regimes beside the MNs' lateral surfaces where superconductivity is strongly suppressed. A direct consequence is that in the case of curve $A^{/}$ the measured resistance should be higher when compared to the one obtained in curves A and C. Notice that this argument fits nicely to the discussion made above and suggests a self consistent explanation for the enhancement of superconductivity observed in figures 9(a)-(b) for curves A and C when compared to curve $A^{/}$. As the respective top view of the HS uncovers while measuring curves A and C the *isolated* regimes that are placed above the MNs have higher total magnetic field $\mathbf{H}_{tot}$ when compared to the externally applied $\mathbf{H}_{ex}$. Thus in these cases the **isolated** regimes are the ones where superconductivity is *suppressed*. Consequently during a transport experiment where curves A and C are detected these areas are short-circuited by the respective **continuous** regimes where superconductivity is *enhanced*.

Finally, we should note that since the internal fields of MNs are nonuniform the modu-

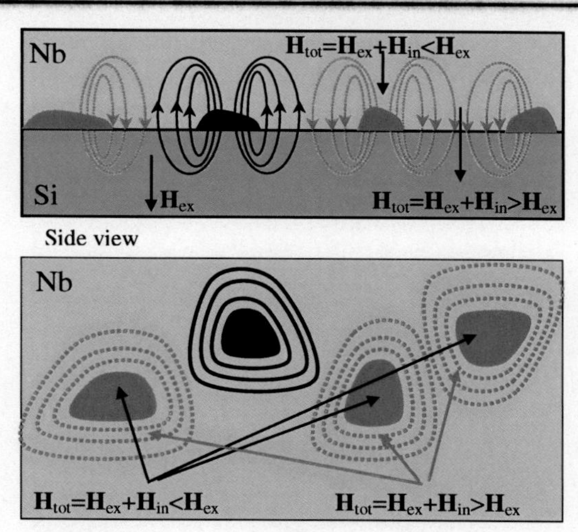

Figure 16: Schematic representation of side (upper panel) and top (lower panel) view of the experimental conditions during the measurement of curve $A^{/}$. Notice that the regimes of the superconductor that are placed above the MNs experience lower total magnetic field $\mathbf{H}_{tot}$ when compared to the externally applied $\mathbf{H}_{ex}$. In contrast, in the regimes by the lateral surfaces the total field $\mathbf{H}_{tot}$ is higher than the external $\mathbf{H}_{ex}$.

lation of the external field is also nonuniform [23, 24, 25, 26]. The compensation of external magnetic field is more pronounced by the lateral surfaces of MNs. Since the superconducting order parameter is enhanced only near the lateral superconductor-ferromagnet interfaces the situation is analogous to surface superconductivity occurring at a superconductor-insulator interface and may be called as magnetically induced surface-like superconductivity.

It should also be noted that magnetic memory effects may be obtained not only in resistance data but are observed also in SQUID magnetization measurements. Representative isofield $m(T)$ data are shown in figures 17 (a) and (b). In the upper panel we focus on the low-field part of the magnetization loop which was obtained for the HS at $T = 9$ K$> T_c$, while the lower panel presents the respective data obtained progressively at points $A, C, A^{/}, C^{/}$ and $A^{//}$. Notice that during the measurement of all these curves the external magnetic field was $|H| = 9$ kOe. The experimental protocol during these measurements is as following. Initially the HS is carefully demagnetized at $T = 10$ K$> T_c$. Then at $H = 0$ Oe the temperature is lowered at the desired value and the magnetic field $H = 9$ kOe is applied. The magnetization of the HS is recorded as a function of temperature until T_c is exceeded. The obtained $m(T)$ curve is the one labelled as curve A (full rhombi in the lower panel). Then the magnetic field is switched off and the temperature is decreased to the desired value for the new measurement. The magnetic field is increased to $H = 55$ kOe and then decreased to $H = 9$ kOe. During this procedure all the MNs are aligned so that although the field is finally lowered to $H = 9$ kOe the magnetic constituent is in its remanent state (point C in the upper panel). The new $m(T)$ curve is recorded while increasing the temperature until T_c is exceeded. The obtained $m(T)$ curve is the one labelled as curve C

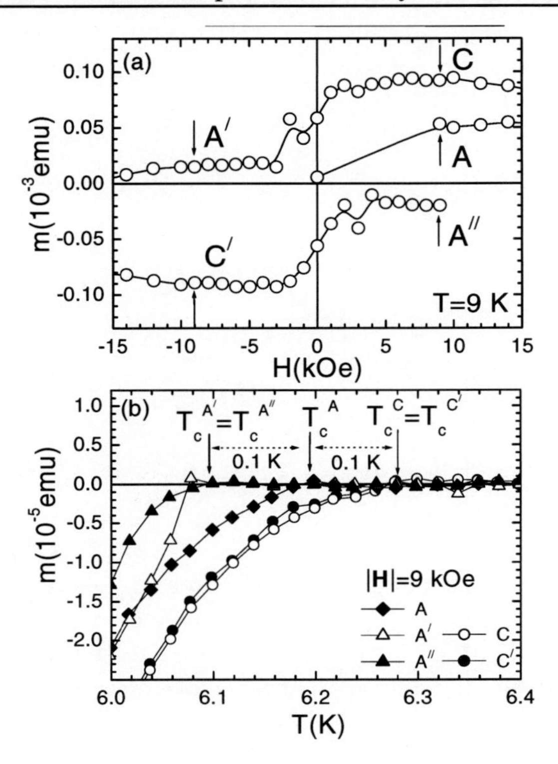

Figure 17: (Upper panel) Typical hysteresis loop of the HS obtained at $T = 9$ K. (Lower panel) Representative magnetization curves $m(T)$ obtained at $|H| = 9$ kOe as a function of temperature for different magnetic histories of the HS. We clearly observe that the curves obtained at the symmetric points $A^{/} - A^{//}$ and $C - C^{/}$ of the magnetization loop exhibit the same upper-critical temperature $T_c(9$ kOe$)$. The curve which was obtained at point A of the magnetization loop (the MN was initially demagnetized) is placed between the respective curves $A^{/} - A^{//}$ and $C - C^{/}$. The applied magnetic field was parallel to the easy-axis $\hat{a}_e$ of magnetization ($\mathbf{H} \parallel \hat{a}_e$).

(open circles in the lower panel). By using the same protocol we performed all other measurements at points $A^{/}, C^{/}$ and $A^{//}$. Note that at points $A^{/}$ and $C^{/}$ the magnetic field is negative in respect to the rest points. Despite that during all measurements the external field was set at $|H| = 9$ kOe from the data presented in figure 17(b) we clearly see that the critical temperature is strongly affected by the magnetic history of the MNs. More specifically, the curves obtained at the symmetric points $A^{/} - A^{//}$ and $C - C^{/}$ of the magnetization loop exhibit the same upper-critical temperature $T_c(9$ kOe$)$. The curve A (see lower panel) which was obtained at point A (see upper panel) of the magnetization loop (the MN was initially demagnetized) is placed between the respective curves $A^{/} - A^{//}$ and $C - C^{/}$. All these magnetization data are in nice agreement to the resistance ones presented above and may be explained with the same arguments. For the second category of HSs the modulation of superconductivity and the memory effects observed in our data may be explained in a consistent way [19, 23, 24, 25, 35, 26].

4.3 Bulk upper-Critical Field Line $H_{c2}(T)$

By taking into account the results presented above it is now easy to understand the behaviour of the bulk upper-critical field $H_{c2}(T)$ and the surface critical field $H_{c3}(T)$ lines of the HS presented in figures 8 (a)-(b). We will refer only to the first category of HSs studied in the present chapter. Let us start with the $H_{c2}(T)$ line. For convenience, in figure 18 we present comparatively the respective lines $H_{c2}^{HS}(T)$ and $H_{c2}^{Nb}(T)$ of the HS and of the reference Nb film in reduced temperature units for clarity. The aim of its inset is to present the *direct qualitative* and *indirect quantitative* agreement of the difference $\Delta H_{c2} = H_{c2}^{HS}(T) - H_{c2}^{Nb}(T)$ of the bulk upper-critical field lines (open circles) with the magnetization loop $4\pi M^{MN}$ of MNs (filled circles) in the respective regime. We should note that we don't have definite information for the two different kinds of MNs (in accordance to CoPt, $CoPt_3$ particles, which are softer than CoPt ones, could also be present) and their relative volume. Thus the respective magnetization curve $4\pi M^{MN}$ can't be calculated directly. Nevertheless, it may be estimated indirectly from the self-consistent comparison of the experimental data with theoretical relations (see the discussion below in the same subsection). The dotted line represents schematically the contribution of the HMNs at low external fields. In general we expect that as the external field increases the number of SMNs that align their dipolar fields also increases. As a result, the compensation of the external field by the dipolar fields of the SMNs in the regimes by their lateral surfaces is getting more efficient and gradually expands in the whole film. Thus, as the external field increases (but still $H < H_{sat}^{soft} \simeq 8$ kOe) superconductivity in the HS should be observed at progressively higher temperatures. This means that the $H_{c2}^{HS}(T)$ line of the HS should gradually diverge from the respective line $H_{c2}^{Nb}(T)$ of the reference Nb film. When all the SMNs are saturated (at $H = H_{sat}^{soft} \simeq 8$ kOe) the additional external field is not further compensated and as a result for $H_{sat}^{soft} \simeq 8$ kOe $< H < H_{onset}^{hard} \simeq 20$ kOe the $H_{c2}^{HS}(T)$ and $H_{c2}^{Nb}(T)$ lines should be almost parallel. The behaviour discussed above is clearly observed in the inset of figure 18. We see that the difference $\Delta H_{c2} = H_{c2}^{HS}(T) - H_{c2}^{Nb}(T)$ traces the magnetization loop up to the onset field of the HMNs. The fact that the magnetization loop and the resulted ΔH_{c2} curve are not absolutely horizontal in the field regime $H_{sat}^{soft} < H < H_{onset}^{hard}$ is due to the contribution of the HMNs as this is qualitatively represented by the dotted line. Furthermore, in the main panel of figure 18 we see that when the onset field of the HMNs is exceeded the line $H_{c2}^{HS}(T)$ of the HS exhibits a strong change in its slope and tends to meet the respective line $H_{c2}^{Nb}(T)$ of pure Nb. This is illustrated in the inset by the sharp drop of their difference ΔH_{c2} above the onset field $H_{onset}^{hard} \simeq 20$ kOe. These results suggest that while the SMNs slightly promote bulk superconductivity the HMNs strongly suppress it.

In a recent paper A.Yu. Aladyshkin et al.[34] have studied the influence of electromagnetic mechanism on the nucleation of superconductivity in a HS consisting of a thin superconducting layer placed on top of a ferromagnetic film of thickness D. In that work the authors assumed that the magnetic domains exhibit a magnetization easy-axis $\hat{a}_e$ normal to the surface of the HS. In addition they assumed that the domain walls are well pinned so that the existing domains, of width w, don't change significantly as a function of an external field applied normal to the surface of the HS so that the modulation of superconductivity depends exclusively on the competition between the robust internal magnetic fields produced by the domains and the external applied field [34]. By assuming that in our case the

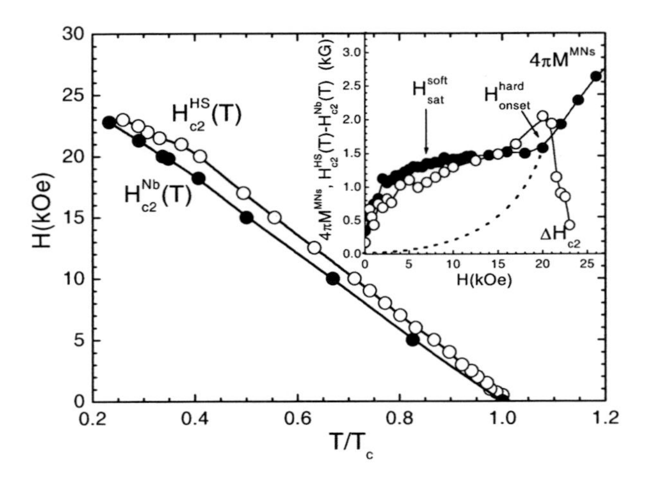

Figure 18: In the main panel are presented the bulk upper-critical fields of the HS (open circles) and pure Nb (filled circles) in reduced temperature units. The $H_{c2}^{\mathrm{Nb}}(T)$ of pure Nb is a straight line in the regime investigated here. Despite that the critical temperatures T_c of the two films are almost the same (8.4 K and 8.6 K for HS and pure Nb respectively) it is necessary to use reduced temperature units in order to reveal the small difference of the two curves observed in the low field regime. The inset presents comparatively the initial part of the magnetization loop $4\pi M^{MN}$ of the MNs (filled circles) and the difference $\Delta H_{c2} = H_{c2}^{\mathrm{HS}}(T) - H_{c2}^{\mathrm{Nb}}(T)$ (open circles) of the two curves presented in the main panel. We note that the $4\pi M^{MN}$ curve is estimated indirectly from the comparison of our experimental data with the theoretical relations (see text). The dotted line represents the contribution of the HMNs at low fields. The applied magnetic field was parallel to the easy-axis $\hat{a}_e$ of magnetization ($\mathbf{H} \parallel \hat{a}_e$).

lateral surfaces of each MNs corresponds to a domain wall (that is, each MNs corresponds to a magnetic domain) the theoretical results obtained in Ref.[34] could be applied to our experimental data. *Of course a noticeable difference is that in our case the dipolar fields of the MNs are not robust but are aligned under the action of the external field.* This has a clear consequence on our experimental results as discussed below. Two limiting situations were considered analytically in Ref. [34]. The case of a thick ferromagnetic layer with $D >> w$ and the opposite one where $w >> D$. The profiles of the internal fields produced by the domains are completely different in these two opposite cases. These internal fields are mainly directed along the easy-axis $\hat{a}_e$, normal to the surface of the HS, and present significant modulation mainly above the domain walls where their normal component gradually becomes zero as it changes direction. In the first case where $D >> w$ the internal fields are almost homogeneous inside the domains and exhibit significant change only near domain walls. As a result in zero external field superconductivity should be promoted mainly above domains walls where the normal component of the internal fields becomes almost zero, while under the presence of an external field the superconducting nucleus should be progressively shifted toward the center of the domains which exhibit magnetization oppositely directed to the external field. [34] In the other limiting case where $w >> D$ the internal fields are mainly restricted very close to the domain walls so that at the center of the domains they are almost zero. In addition, as discussed above near domain walls the component of the internal fields which is normal to the surface of the HS is negligible due to the change in its direction between the adjacent magnetic domains. Thus, in this limiting case under zero external field superconductivity could be preserved almost throughout the whole superconducting film. Since for $w >> D$ the superconducting properties are maintained almost in the whole superconductor this limiting case is more favorable for the possible observation of bulk superconductivity. Due to the effective zeroing of the internal fields inside the domains the superconducting nucleus should be pronounced in these regimes comparing to the regimes of the superconductor placed above domain walls. But as the external field is gradually increased the orbital effect suppresses the superconducting nucleus inside the domains (where the internal fields are almost zero and can't compensate the external field). Then superconductivity may be favorably observed mainly in the regimes near domain walls where the oppositely directed internal fields may compensate efficiently the external field.

Since in our case the MNs have a flat shape with their in-plane dimensions $w = 50-500$ nm much higher than their thickness $D = 10 - 50$ nm we may compare our experimental results with the theoretical outcome of Ref.[34] for the second limiting case $w >> D$. In this case the theoretical results imply that the upper-critical field line $H_{c2}(T)$ should exhibit a change in the slope when the external field becomes almost equal to the saturation magnetization $4\pi M_{sat}$ of the MNs. In high fields ($H > 2\pi M_{sat}$) $H_{c2}(T)$ exhibits lower slope than its low-field ($H < 2\pi M_{sat}$) part. Qualitatively, the same feature is observed in our experimental $H_{c2}(T)$ data presented in detail at the inset of the lower panel of figure 8. Despite the qualitative agreement it should be noted that in our case the change in the slope of $H_{c2}(T)$ is observed at $H = H_{sat}^{soft} \simeq 8$ kOe, that is, when the applied field becomes equal to the saturation field H_{sat}^{soft} of the SMNs (the other change in the slope of $H_{c2}(T)$ observed at $H = H_{onset}^{hard} \simeq 20$ kOe will be discussed below). We believe that this is a practical condition needed for the efficient compensation of the external field and the observation of the respective effect, since in our HS the dipolar fields of the MNs *are not robust (as in the theoretical treatment) but are gradually aligned under the influence of the external field.* Quantitatively at very low ($H << 2\pi M_{sat}$) and high ($H > 2\pi M_{sat}$) external fields the upper-critical field line $H_{c2}^{HS}(T)$ of the HS may be approximated by the respective theoretical expressions[34]

$$H_{c2}^{HS}(T) \propto \sqrt{2\pi M_{sat}}(1 - \frac{T}{T_{co}})^{3/4} \tag{1}$$

and

$$\frac{T - T_{co}}{\Delta T_c^{orb}} = 1 - \frac{H_{c2}^{HS}(T)}{2\pi M_{sat}}, \tag{2}$$

where T_{co} is the zero-field critical temperature of the superconductor, $\Delta T_c^{orb} = \frac{T_{co}}{2}\frac{2\pi\xi_o^2}{\Phi_o}4\pi M_{sat}$ is the shift of the critical temperature due to the influence of the orbital effect owing to the internal fields of the MNs and finally $4\pi M_{sat}$ is the saturation magnetization of the MNs [34]. We also remind that ξ_o is the coherence length at zero temperature, which in our case is $\xi_o \simeq 100$ Å. By substituting ΔT_c^{orb} Eq.2 may be rewritten as

$$H_{c2}^{HS}(T) = -\frac{\Phi_o}{2\pi\xi_o^2 T_{co}}T + \frac{\Phi_o}{2\pi\xi_o^2} + 2\pi M_{sat}. \quad (3)$$

For a pure superconductor (in our case Nb) the respective relation for the bulk upper-critical field is

$$H_{c2}^{Nb}(T) = -\frac{\Phi_o}{2\pi\xi_o^2 T_{co}}T + \frac{\Phi_o}{2\pi\xi_o^2}. \quad (4)$$

According to the theoretical results of Ref.[34] we see that for magnetic fields $H << 2\pi M_{sat}$ (that is, very close to T_c) an unusual $H_{c2}^{HS}(T) \propto (1 - \frac{T}{T_{co}})^{3/4}$ temperature dependence should be observed (relation 1), while for fields $H > 2\pi M_{sat}$ the bulk upper-critical field line $H_{c2}^{HS}(T)$ of the HS (relation 3) is simply *parallel* to the respective line $H_{c2}^{Nb}(T)$ of the pure superconducting layer (relation 4). *Their only difference is that $H_{c2}^{HS}(T)$ is shifted to higher magnetic fields by an amount of $2\pi M_{sat}$.* Although we didn't observe any anomalous dependence in our $H_{c2}^{HS}(T)$ data close to T_c, at higher external fields our experimental results agree fairly with the theoretical expectations. This is clearly revealed in the data presented in figure 18. As we noticed in the discussion given above for $H > H_{sat}^{soft} \simeq 8$ kOe the upper-critical field lines of HS and of pure Nb are almost parallel since their difference traces the magnetization of the SMNs (see the inset of figure 18). In addition to the qualitative agreement we also performed a quantitative comparison by using Eq. 3 to describe our experimental results. As we see in the inset of the lower panel of figure 8 the extrapolation of the middle-field part to zero field implies that the shift of the critical temperature is $\Delta T_c^{orb} \simeq 0.2$ K. From this quantity the saturation magnetization of the MNs could be estimated by using the relation $\Delta T_c^{orb} = \frac{T_{co}}{2}\frac{2\pi\xi_o^2}{\Phi_o}4\pi M_{sat}$. The resulted value is $4\pi M_{sat} \simeq 1500$ Oe which is in fair agreement to the value exhibited by curve $\Delta H_{c2} = H_{c2}^{HS}(T) - H_{c2}^{Nb}(T)$ in the respective regime of magnetic fields (see inset of figure 18). We see that the comparison of experimental data with the theoretical suggestions leads to consistent results. Furthermore, according to Eq. 3 the zero-temperature value of bulk upper-critical field is $H_{c2}^{HS}(T = 0) = \frac{\Phi_o}{2\pi\xi_o^2} + 2\pi M_{sat} \simeq 32.6$ kOe. The respective value obtained by extrapolation of the experimental data down to $T = 0$ is $H_{c2}^{HS}(T = 0) = 33.5$ kOe (see lower panel of figure 8). We observe that the obtained values are consisted with each other, as should be expected. Regarding the bulk upper-critical field line $H_{c2}(T)$ we conclude that the effect observed in our experimental results at the saturation field $H_{sat}^{soft} \simeq 8$ kOe of the SMNs may be explained according to the theoretical proposals of Ref.[34]. The underlying physical situation described by Eqs. 1 and 2 is that as we progressively increase the external field the superconducting nucleus gradually shifts from the regimes of the superconductor which are placed well inside magnetic domains to regimes that are placed near domain walls. *The respective process occurring in our case is that for $H < H_{sat}^{soft} \simeq 8$ kOe the superconducting order parameter obtains high values in the regimes away from the lateral surfaces of the SMNs, while for $H > H_{sat}^{soft} \simeq 8$ kOe superconductivity is promoted near the lateral surfaces of SMNs*.

On the other hand at the onset field $H_{onset}^{hard} \simeq 20$ kOe of the HMNs the $H_{c2}(T)$ data present an additional change in the slope which is more abrupt. Since the feature observed at $H = H_{onset}^{hard} \simeq 20$ kOe is qualitatively the same with the one observed at the saturation field $H_{sat}^{soft} \simeq 8$ kOe of the SMNs we could assume that the same explanation may also be applied. Unfortunately, a comparison of the experimental results with the theoretical expression given by Eq.3 resulted in a large quantitative disagreement. Thus, we believe that the mechanism leading to the modulation of bulk superconductivity in the high-field regime $H > H_{\rm onset}^{\rm hard} \simeq 20$ kOe is different from the one that we used to describe the low-field $H < H_{\rm sat}^{\rm soft} \simeq 8$ kOe data. A possible explanation could rely on the fact that the increased exchange field of the HMNs is strong enough to induce pronounced pair-breaking in the bulk of the superconductor due to the proximity effect [44, 45, 46, 47]. As a result at high fields superconductivity is strongly suppressed.

4.4 Surface-Superconductivity Critical-Field Line $H_{c3}(T)$

The behaviour of the normal state boundary $H_{c3}(T)$ presented in figure 8 (a) may be explained in a consistent way by using the same arguments evolved above. In the low-field regime $H < H_{\rm sat}^{\rm soft} \simeq 8$ kOe the SMNs are not entirely aligned and as a result the compensation of the external field by their dipolar fields is not complete in the whole film. In contrast, when the saturation field $H_{\rm sat}^{\rm soft} \simeq 8$ kOe is reached, the magnetization of the SMNs (which we remind that is opposite to the external field in the regime by their lateral surfaces) takes a value high enough so that compensation is now efficient and takes place in the whole film's area. Since the discussed effect takes place mainly when all the SMNs are aligned we speculate that at microscopic level the whole process relies on the formation of percolation paths that trace the lateral surfaces of the SMNs in the whole film's area. These paths assist the superconducting component of the transport current and as a result lower values of the resistance are maintained at higher temperatures. Of course, such a percolation assisted mechanism could be straightforwardly clarified by a study where the coverage of the surface with particles is varied systematically from zero coverage (no particles) to total coverage (continuous film). Reasonably, the effect under discussion should be pronounced at some specific value of the density of particles and will be absent in the limiting cases of zero and full coverage.

In a similar way for $H > H_{\rm sat}^{\rm hard} \simeq 36$ kOe the $H_{c3}(T)$ line exhibits another more abrupt upturn. As a result, the high-field regime where the resistance is lower than the normal state value is greatly enhanced. Unfortunately above the saturation field $H_{\rm sat}^{\rm hard} \simeq 36$ kOe all the HMNs are saturated so that their ability to compensate the external magnetic field is entirely expended. Thus, as the applied field still increases gradually weakens the superconducting component of the transport current and as a result the resistance increases. This is the reason why the voltage curves presented at figure 6 (b) gradually broaden and tend to become horizontal as the external field increases above $H_{\rm sat}^{\rm hard} \simeq 36$ kOe.

Regarding $H_{c3}(T)$ our experimental data are in nice agreement to recent theoretical results as discussed below. In a recent publication Sa-Lin Cheng and H.A. Fertig [35] have studied how $H_{c3}(T)$ is modulated under the influence of an external magnetic field in a basic HS that consists of a magnetic dot placed at the center of a superconducting film. It is well known that for the case of a superconductor-insulator bilayer the superconducting order parameter must exhibit a vanishing derivative (zero normal component of the supercurrent) at their interface. As a result, when a magnetic field is applied, the states that satisfy the linearized Ginzburg-Landau equation and this specific boundary condition exhibit a maximum at the surfaces of the superconductor, leading to a situation where the superconducting

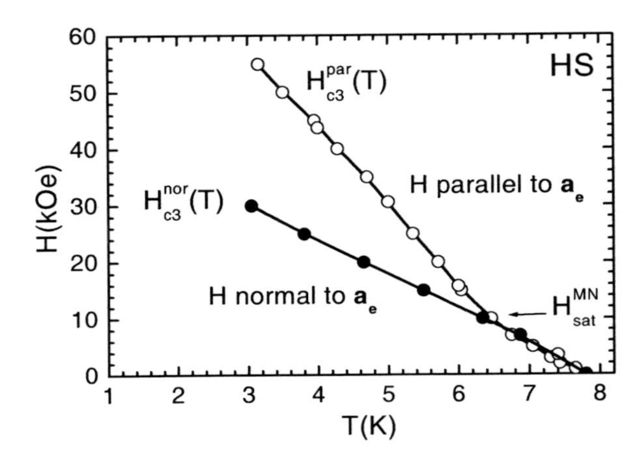

Figure 19: Comparative presentation of the surface superconductivity field lines $H_{c3}^{HS}(T)$ for a simple HS when the field is applied parallel (open circles) and normal (solid circles) to the easy-axis $\hat{a}_e$ of magnetization of MNs.

order parameter is nonzero at the interface, while it vanishes in the bulk of the specimen. When a magnetic dot is introduced, the strong exchange field strongly reduces the superconducting order parameter *inside* the ferromagnet. In such situations it is appropriate to adopt a boundary condition in which the order parameter vanishes at the superconductor-ferromagnet interface. Thus it is generally expected that surface superconductivity should not be observed in such systems. Sa-Lin Cheng and H.A. Fertig [35] stressed that the problem of a ferromagnetic dot hosted in a superconducting film has an important parameter that had not been addressed in previous theoretical investigations: the magnetization and the related stray magnetic field of the dot itself. Their work gave evidence that, indeed, magnetic dots support and can presumably "pin" multiple vortices. They showed [35] that the enhancement of $H_{c3}(T)$ depends on the saturation magnetization and on the geometric characteristics of the employed dot.

The HS studied in our work is similar to the model film studied in Ref. [35] and also the exact experimental configuration is the same. Thus a comparison of our experimental results to the theoretical outcome of Ref.[35] may be done. The results of Ref.[35] revealed that the internal dipolar field of the ferromagnetic dot compensates the external magnetic field in the region very near to its lateral surface. As a result the critical field $H_{c3}(T)$ is enhanced and depending on the saturation magnetization and on the geometric characteristics of the dot it may significantly exceed the value $H_{c3}(T) = 1.695 \cdot H_{c2}(T)$ which holds for the case of a superconducting-insulating boundary [39]. More specifically, the authors showed [35] that the surface critical field $H_{c3}(T)$ increases as (i) the saturation magnetization and (ii) the aspect ratio thickness/diameter of the dot increase. Regarding the dependence of $H_{c3}(T)$ on the saturation magnetization of the ferromagnetic dot our experimental results presented above nicely agree with the theoretical predictions of Ref.[35]. Furthermore, in the model structure studied in Ref.[35] the aspect ratio thickness/diameter of the ferromagnetic dot determines its magnetic anisotropy (easy-axis $\hat{a}_e$ of magnetization). It was proposed that $H_{c3}(T)$ should be strongly enhanced when the thickness of the dot becomes equal to its diameter and as a result the easy-axis $\hat{a}_e$ of magnetization is normal to the film's surface [35]. In our case the origin of the magnetic anisotropy exhibited by the MNs is more complicate since it is not determined only by their geometric characteristics. Despite that, their easy-axis $\hat{a}_e$ of magnetization is normal to the surface of the HS. In agreement to the theoretical results of Ref. [35] our experimental results show enhancement of $H_{c3}(T)$ only for the case where the external field was parallel to the easy-axis $\hat{a}_e$ of magnetization (normal to the surface of the HS). When the external field was applied normal to $\hat{a}_e$ (parallel to the surface of the HS) no significant effect was observed in $H_{c3}(T)$. The respective comparative data are shown in figure 19. These data show clearly that the MNs don't affect the superconducting properties for the case where the external field is applied along their hard magnetization axis.

5 Possible Applications of HS

In recent years, ferromagnetic-superconducting HS have attracted the interest of the scientific community not only due to their importance for the investigation of basic physics but also due to various practical applications that generally could be based on hybrid structures. It was D. Givord, Y. Otani and coworkers who first introduced a HS consisting of a Nb layer in close proximity to (but electrically isolated from) an *ordered array* of MNs and noted the possible application of the MNs for the controlled modulation of the superconducting order parameter [14, 15]. Since then many groups investigated similar HSs by performing extensive magnetic and mainly magnetoresistance measurements especially close to the zero-field critical temperature [16, 17, 18, 19, 20, 21, 22]. The possible applications of HS could be categorized into two wide classes: (a) large scale devices for power applications where increased current-carrying capability is the main objective (needed, for instance, for the operation of resistive-superconductive hybrid magnets required for the production of high magnetic fields [50]) and (b) small scale devices where increased sensitivity is the desired quantity (needed for instance for the operation of magnetoresistive storage devices and spin valves, [51, 27, 28, 29, 49, 25] and lenses of magnetic flux [52].) In our work we focus on the study of hybrid structures which consist of *randomly distributed* MNs embedded in an Nb layer since the preparation of such HS is much more easier than the one of *ordered arrays*. Thus, the *randomly distributed* MNs are more attractive for commercial applications. Despite that, some of the discussed physical phenomena should equally be observed in both types of HS.

As discussed above, in the first category of applications the current-carrying capability is the main goal. Thus, the main objective is the enhancement of the part of the $H-T$ operational diagram where the superconductor maintains the maximum of its superconducting properties [23, 25, 26]. Since the compensation ability of the MNs on microscopic level depends on the strength of their dipolar fields (on macroscopic level on their saturation magnetization) by inspection of the data presented in figures 4 (a) and 8 (a) we conclude that the slope of the $H_{c3}(T)$ line is determined by the saturation magnetization of the MNs. In addition, figure 8 (a) shows that the high-field part of the $H_{c3}(T)$ line exhibits a two-fold (threefold) change in its slope in comparison to the middle-field (low-field) part. Thus, Θ (3Θ) is the angle between the middle-field (high-field) and low-field $H_{c3}(T)$ data (2Θ is the respective angle between the high-field and middle-field parts). In accordance, in figure 4

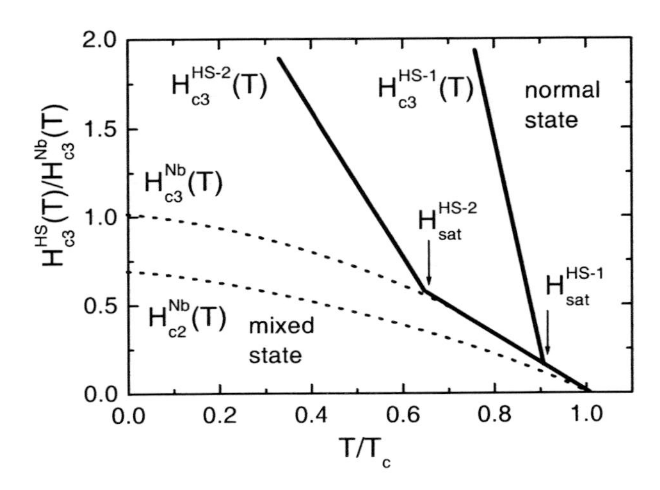

Figure 20: Qualitative comparative presentation of the surface superconductivity field lines $H_{c3}^{\mathrm{Nb}}(T)$ and $H_{c3}^{\mathrm{HS}}(T)$ of pure Nb and two hybrid films. The first HS exhibits lower saturation field comparing to the second one, $H_{\mathrm{sat}}^{\mathrm{HS-1}} < H_{\mathrm{sat}}^{\mathrm{HS-2}}$. In addition, the saturation magnetization of the MNs in the first HS is higher than the one of the second. As a result the $H_{c3}^{\mathrm{HS-1}}(T)$ line of the first HS is placed at fields and temperatures much higher comparing to the respective lines $H_{c3}^{\mathrm{HS-2}}(T)$ of the second HS and $H_{c3}^{\mathrm{Nb}}(T)$ of pure Nb. The applied magnetic field is assumed to be parallel to the easy-axis $\hat{a}_e$ of magnetization of the MNs $(\mathbf{H} \parallel \hat{a}_e)$.

(a) we see that the saturation magnetization of the HMNs is two times the respective value of the SMNs, $m_{\mathrm{sat}}^{\mathrm{hard}} \simeq 2m_{\mathrm{sat}}^{\mathrm{soft}}$. Thus we conclude that the change in slope of the $H_{c3}(T)$ line is determined by the saturation magnetization of the embedded particles. In addition, the maximum enhancement of the operational $H - T$ regime is achieved when the steep increase of the $H_{c3}(T)$ line is initiated from the lowest possible field. Thus we propose that MNs with high saturation magnetization but of low saturation fields could enhance dramatically the regime where a HS could be useful for current-carrying applications. The above argument is schematically presented in figure 20 for the case of particles that exhibit single magnetic behavior. We have cross-checked this result by producing HSs that host particles that exhibit a singe magnetic phase. We note that this figure presents qualitatively the ideal single-magnetic-phase particles wanted for our purposes.

Going a step further we propose that in a comprehensively designed HS the transport supercurrent could be selectively assisted by the upper or the lower part of its thick Nb layer in the following way: at external magnetic fields much lower than the bulk upper-critical field $H < H_{c2}^{\mathrm{Nb}}(T)$ it is logical to assume that the transport supercurrent is mainly supported by the bulk upper part of the Nb layer. This upper part is not affected by the internal fields of the MNs due to their limited range (of the order of a few decades of nm). On the other hand recall that at zero external field the lower part of the Nb layer is subjected to the internal fields of the MNs which since they are aligned they strongly suppress superconductivity in the regime under discussion. It is only the increase of the external field that restores superconductivity in the lower part of the Nb layer in the range

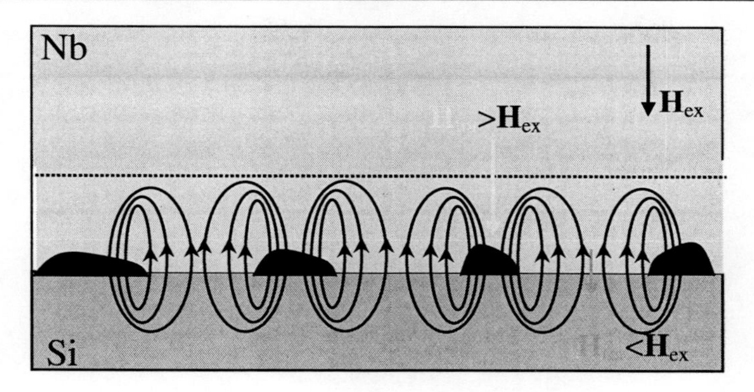

Figure 21: Schematic presentation of a HS appropriate for current-currying applications. The MNs are in the remanent state. The upper part of the Nb layer is not affected by their dipolar fields since is placed outside the finite range indicated by the horizontal dot line. The lower part is strongly influenced by the MNs' dipolar fields experiencing higher (lower) total magnetic field in the regimes above the MNs (by the lateral surfaces of MNs). As a result when the bulk upper-critical field is exceeded the only part of the HS appropriate to promote an applied current is its lower part.

of fields $H < H_{\rm sat}^{\rm MNs}$ as it gradually compensates the internal dipolar fields of MNs. The situation is completely different in the regime of high magnetic fields. When the external field exceeds the bulk upper-critical field, $H > H_{\rm c2}^{\rm Nb}(T)$ the upper Nb part can no longer support the transport supercurrent adequately since it enters the normal metallic state. On the other hand according to our experimental results presented above the lower part of the Nb layer should experience a total magnetic field which is lower than the external applied field thus this lower part preserves its superconducting properties at higher external fields when compared to the upper part of the layer. Of course we should keep in mind that since the effective compensation of the external field is limited only very close to the lateral surfaces of the MNs the lower part of the Nb layer preserves the desired properties only in the range of fields $H_{\rm c2}(T) < H < H_{\rm c3}(T)$. Thus, given that $H > H_{\rm sat}^{\rm HS}$, in the range of external fields under discussion ($H > H_{\rm c2}^{\rm Nb}(T)$) the lower part of the Nb layer could mostly support the transport supercurrent until the surface superconductivity field $H_{\rm c3}^{\rm HS}(T)$ is exceeded. The whole idea is schematically presented in figure 21. The horizontal dot line indicates the vertical range of the MNs' dipolar fields. Since the upper part of the Nb layer is not affected by the internal fields it exhibits the usual behavior of a single superconducting film. In contrast, the lower part which hosts the MNs in the high-field regime should exhibit enhanced surface superconducting properties when compared to the upper part and should preserve the applied transport supercurrent when the external field has exceeded the bulk upper-critical field $H_{\rm c2}^{\rm Nb}(T)$.

The second wide class of applications relies on the change of a HS's resistance under the application of a magnetic field. Under this sense a ferromagnetic-superconducting HS could be a promising competitor of the giant magnetoresistance elemental devices [51, 27, 28, 29, 49, 24, 25, 26]. Recently, great progress has been reported in various experimental works where the magnetoresistance of comprehensive HS devices has been studied. J.Y. Gu et al. reported a 25% change in the measured resistance which was motivated by

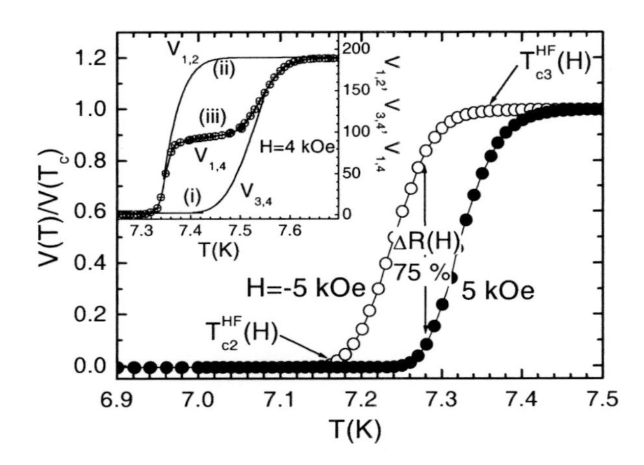

Figure 22: Normalized voltage measured in a single HS at $H = \pm 5$ kOe. The observed shift due to the change in the direction of the external field is $\Delta V(\pm 5kOe) = 75\%$. The inset presents schematically the three states of an elemental device which could be based on our CS. In all cases the magnetic field is parallel to the magnetization easy-axis $\hat{a}_e$ of the MNs ($\mathbf{H} \parallel \hat{a}_e$).

the relative orientation of the magnetization of the magnetic layers in a ferromagnetic-superconducting-ferromagnetic trilayer structure. [29] Soon after, A. Yu. Rusanov et al. showed that in a more simple ferromagnetic-superconducting bilayer the breaking of Cooper pairs can be significantly suppressed in the regimes of domain walls [49]. The so-called "enhancement of T_c induced by domain walls" resulted in a clear modulation of the magnetoresistance curves ($\Delta T_c \simeq 10$ mK). It should be noted that in the works mentioned above it was mainly the selective control of the exchange field that allowed the modulation of magnetoresistance. In our case the situation is more simple. The modulation of magnetoresistance is achieved by the selective control of the stray fields of the MNs as discussed above. Depending on the direction of the applied field we obtained a 75% change in the value of magnetoresistance and a shift of the isothermal $V(H)$ curves $\Delta H \simeq 400$ Oe. As discussed above the respective temperature shift of the isofield $V(T)$ curves is of the order $\Delta T_c \simeq 100$ mK.

The data presented in figure 22 were obtained in a simple HS when the external field was changed from $H = +5$ kOe to $H = -5$ kOe. A pronounced percentage change $\Delta V(\pm 5kOe) = 75\%$ is observed in the measured voltage which is motivated by the magnetic state of the MNs. Thus, a simple HS could probably serve as a convenient *bistate* magnetoresistive memory device. Furthermore, a CS could serve as a *tristate* memory device which, at constant temperature, exhibits three distinct resistive states as is schematically presented in the inset of Fig.22: (i) zero resistance (output at $V_{3,4}(T)$), (ii) normal state resistance (output at $V_{1,2}(T)$) and (iii) an intermediate value of the resistance (output at $V_{1,4}(T)$). Interestingly, the CS can be tuned between states (i), (ii) and (ii) under the application of an external field and more importantly its zero-resistance state (i) and intermediate state (iii) can be additionally modulated by the specific magnetic state of the MNs. This device could be considered as a development of the recently proposed magnetoresistive memory and superconducting spin valve bistate elements [27, 28, 29]. It should be noted that while the operation of those bistate elements [27, 28, 29] is based on the exchange-bias mechanism our tristate elemental device simply works under the action of the dipolar fields of the MNs as discussed in the preceding Sections of this chapter.

6 Conclusion

In conclusion, we presented detailed magnetic and transport data for two categories of HSs consisting of CoPt MNs embedded in high-quality thick Nb layers. We found out that the properties of the superconductor are strongly altered since they depend on both the saturation magnetization and the saturation field of the MNs. The pronounced memory effects observed in the superonducting properties of Nb could be only attributed to the influence of the internal fields produced by the MNs. The constructed phase diagram of the HS presents strong differences when compared to the respective one of the pure superconductor. More specifically, the bulk upper-critical field line $H_{c2}(T)$ is increased by the soft MNs but is strongly reduced by the hard ones. In contrast, surface superconductivity is promoted by the MNs. The respective line $H_{c3}(T)$ exhibits a strong upturn at the saturation field of the embedded particles. Above the field where the upturn occurs its slope is determined by the saturation magnetization of the employed MNs. As a result the local compensation of the external field by the dipolar fields of the MNs greatly enhances the high-field regime where a host superconductor could be used for practical applications. We believe that the use of MNs that exhibit low saturation fields and high magnetization values could result in a wide $H - T$ regime where a conventional low-T_c superconductor could be more attractive for current-carrying applications. Finally, the controlled modulation of superconductivity by the magnetic component suggests that such HSs could be very important for the design of magnetoresistive memory devices in the near future.

References

[1] Tinkham M 1996 *Introduction to superconductivity* (McGraw-Hill)

[2] Ketterson J B & Song S N 1999 *Superconductivity* (Canbridge University Press)

[3] Cullity B D 1972 *Introduction to magnetic materials* (Addison-Wesley)

[4] Chikazumi S 1966 *Physics of Magnetism* (John Wiley & Sons)

[5] Jaccarino V and Peter M 1962 *Phys. Rev. Lett.* **9** 290

[6] Schwartz B B and Gruenberg L W 1969 *Phys. Rev.* **177** 747

[7] Fischer Ø, Jones H, Bongi G, Frei C and Treyvaud A 1971 *Phys. Rev. Lett.* **26** 305

[8] Lee I J, Chaikin P M and Naughton M J 2000 *Phys. Rev. B* **62** R14669

[9] Pissas M, Stamopoulos D, Lee S and Tajima S 2004 *Phys. Rev. B* **70** 134503

[10] Civale L, Marwick A D, Worthington T K, Kirk M A, Thompson J R, KrusinElbaum L, Sun Y, Clem J R and Holtzberg F 1991 *Phys. Rev. Lett.* **67** 648

[11] Chikumoto N, Konczykowski M, Terai T and Murakami M 2000 *Supercond. Scienc. Technol.* **13** 749

[12] Eisterer M, Zehetmayer M, Tönies S, Weber H W, Kambara M, Hari Babu N, Cardwell D A and Greenwood L R 2002 *Supercond. Scienc. Technol.* **15** L9-L12

[13] PerkinsG G K, Bugoslavsky Y, Caplin A D, Moore J, Tate T J, Gwilliam R, Jun J, Kazakov S M, Karpinski J and Cohen L F 2004 *Supercond. Scienc. Technol.* **17** 232

[14] Geoffroy O, Givord D, Otani Y, Pannetier B and Ossart F 1993 *J. Magn. Magn. Mater.* **121** 223

[15] Otani Y, Pannetier B, Nozières J P and Givord D 1993 *J. Magn. Magn. Mater.* **126** 622

[16] Martin J I, Velez M, Hoffmann A, Schuller Ivan K and Vicent J L 1999 *Phys. Rev. Lett.* **83** 1022

[17] Morgan D J and Ketterson J B 1998 *Phys. Rev. Lett.* **80** 3614

[18] Stoll O M, Montero M I, Guimpel J, Akerman Johan J and Schuller Ivan K 2002 *Phys. Rev. B* **65** 104518

[19] Lange M, Van Bael M J, Bruynseraede Y and Moshchalkov V V 2003 *Phys. Rev. Lett.* **90** 197006

[20] Van Bael M J, Lange M, Raedts S, Moshchalkov V V, Grigorenko A N and Bending S J 2003 *Phys. Rev. B* **68** 14509

[21] Jaque D, Gonzlez E M, Martin J I, Anguita J V and Vicent J L 2002 *Appl. Phys. Lett.* **81** 2851

[22] Yang Z, Lange M, Volodin A, Szymczak R and Moshchalkov V 2004 *Nature Materials* **3** 793

[23] Stamopoulos D, Pissas M, Karanasos E, Niarchos D and Panagiotopoulos I 2004 *Phys. Rev. B* **70** 054512

[24] Stamopoulos D, Manios E, Pissas M and Niarchos D 2004 *Supercond. Sci. Technol* **17** L51-L54

[25] Stamopoulos D, Pissas M and Manios E, 2005 *Phys. Rev. B* **71** 014522

[26] Stamopoulos D, Manios E, Pissas M and Niarchos D 2005 *Supercond. Sci. Technol* **18** 538

[27] Tagirov L R 1999 *Phys. Rev. Lett.* **83** 2058

[28] Buzdin A I, Vedyayev A V and Ryzhanova N V 1999 *Europhys. Lett.* **48** 686

[29] Gu J Y, You C-Y , Jiang J S, Pearson J, Bazaliy Ya B and Bader S D 2002 *Phys. Rev. Lett.* **89** 267001; Gu J Y, You C-Y, Jiang J S and Bader S D 2003 *Jour. Appl. Phys.* **93** 7696

[30] Karanasos V, Panagiotopoulos I, Niarchos D, Okumura H and Hadjipanayis G C 2001 *Appl. Phys. Lett.* **79** 1255

[31] Manios E, Karanasos V, Niarchos D and Panagiotopoulos I 2004 *J. Magn. Magn. Mater.* **272-276** 2169

[32] Stamopoulos D, Speliotis A and Niarchos D 2004 *Supercond. Sci. Technol.* **17** 1261; Stamopoulos D and Niarchos D 2004 *Physica C* **417** 69

[33] Klemmer T J, Liu C, Shukla N, Wu X W, Weller D, Tanase M, Laughlin D E, Soffa W A 2003 *J. Magn. Magn. Mater.* **266** 79

[34] Aladyshkin A Yu, Buzdin A I, Fraerman A A, Mel'nikov A S, Ryzhov D A and Sokolov A V 2003 Phys. Rev. B **68**, 184508

[35] Cheng Sa-Lin and Fertig H A 1999 *Phys. Rev. B* **60** 13107

[36] Hsu J W P and Kapitulnik A 1992 *Phys. Rev. B* **45** 4819

[37] Welp U, Rydh A, Karapetrov G, Kwok W K, Crabtree G W, Marcenat C, Paulius L M, Lyard L, Klein T, Marcus J, Blanchard S, Samuely P, Szabo P, Jansen A G M, Kim K H P, Jung C U, Lee H S, Kang B and Lee S I 2003 *Physica C* **385** 154; Welp U, Rydh A, Karapetrov G, Kwok W K, Crabtree G W, Marcenat C, Paulius L, Klein T, Marcus J, Kim K H P, Jung C U, Lee H S, Kang B and Lee S I 2003 *Phys. Rev. B* **67** 12505

[38] Rydh A, Welp U, Hiller J M, Koshelev A, Kwok W K, Crabtree G W, Kim K H P, Jung C U, Lee H S, Kang B and Lee S I cond-mat/0307445 v1

[39] Saint-James D and Gennes P G 1963 *Physics Letters* **7** 306

[40] DeSorbo W 1964 *Phys. Rev.* **135** A1190

[41] Hikita M, Tajima Y, Tamamura T and Kurihara S 1990 *Phys. Rev. B* **42** 118

[42] Ginzburg V L 1956 Zh. ksp. Teor. Phys. **31** 202 [1956 *Sov. Phys. JETP* **4** 153]

[43] De Gennes P G and Sarma G 1963 *Jour. Appl. Phys.* **34** 1380

[44] Buzdin A I, Bulaevskii L N and Panyukov S V 1984 *Zh. ksp. Teor. Phys.* **87** 299 [1984 *Sov. Phys. JETP* **60** 174]

[45] Buzdin A I and Mel'nikov A S 2003 *Phys. Rev. B* **67** 020503(R)

[46] Aarts J, Geers J M E, Brck E, Golubov A A and Coehoorn R 1997 *Phys. Rev. B* **56** 2779

[47] Baladié I and Buzdin A I 2001 *Phys. Rev. B* **64** 224514

[48] Matthias B T and Suhl H 1960 *Phys. Rev. Lett.* **4** 51

[49] Rusanov A Yu, Hesselberth M, Aarts J and Buzdin A I 2004 *Phys. Rev. Lett.* **93** 057002

[50] Bird M D 2004 *Supercond. Sci. Technol.* **17** R19 and references therein.

[51] Oh S, Youm D and Beasley M R 1997 *Appl. Phys. Lett.* **71** 2376

[52] Villegas J E, Savelév S, Nori F, Gonzalez E M, Anguita J V, Garcia R and Vicent J L 2003 *Science* **302** 1188

In: Superconductivity, Magnetism and Magnets
Editor: Lannie K. Tran, pp. 139-172
ISBN: 1-59454-845-5

Chapter 5

SUPERCONDUCTING AND RESISTIVE TILTED COIL MAGNETS: MAGNETIC AND MECHANICAL ASPECTS

Andrey M. Akhmeteli[1]*, Andrew V. Gavrilin[2]** and W. Scott Marshall[3]*

[1]Microinform Training Center, 38-2-139 Rublevskoe shosse, 121609 Moscow, Russia
[2]National High Magnetic Field Laboratory – Florida State University, Florida 32310, USA
[3]Tai-Yang Research, Inc., 9112 Farrell Park Lane, Knoxville, Tennessee 37922-8525, USA

Abstract

The mathematical foundation is laid for a relatively new type of magnets for generation of uniform transverse magnetic field in a large volume of space - tilted coil magnets. These consist of concentric nested solenoidal coils with elliptical turns tilted at a certain angle to the central axis and current flowing in opposite directions in the coils tilted at opposite angles, generating a perfectly uniform transverse field. Both superconducting wire-wound and resistive Bitter tilted coils are discussed. An elegant analytical method is used to prove that the wire-wound tilted coils have the ideal distribution of the axial current density, "cosine-theta" plus constant. Magnetic fields are calculated for a tilted Bitter coil magnet using an original exact solution for current density in an elliptical Bitter disk. This solution was obtained for disk shapes bounded by homothetic ellipses. A much simpler analytical solution was found for the case of confocal ellipses. A finite element elastic stress analysis was also conducted and revealed rather moderate, i.e., acceptable, stresses in the coils. Winding technologies for tilted coils are discussed. Superconducting wire-wound tilted coil magnets may become an alternative to traditional dipole magnets for accelerators, and Bitter tilted coil magnets are attractive for rotation experiments with a large access port perpendicular to the field and for testing of long samples of superconducting wires and cables in high magnetic fields. It is noteworthy that tilted Bitter magnets represent the only option to test long pieces of high field

* E-mail address: akhmeteli@home.domonet.ru
** E-mail address: gavrilin@magnet.fsu.edu, fax: (850) 644-0867, phone: (850) 645-4906)
*** E-mail address: wsmarshall@earthlink.net, fax: (865) 560-5542, phone: (865) 250-0237)

superconductors and other lengthy materials in transverse uniform fields above 15T, which makes such magnets very attractive from a practical standpoint.

I Introduction

A number of important applications require a high magnetic field perpendicular to the long dimension of the magnet. Such a 'transverse' magnetic field can be obtained with several magnet configurations [1]. Among the most practical 'transverse field' windings, superconducting curved-saddle 'dipoles' (Fig. 1a) have the advantage of providing a large volume of space with a high and uniform transverse field; this is required, e.g., for accelerator magnets. In fact, apart from the end effects, the transverse field is perfectly uniform across the entire aperture of such a dipole, as the winding cross-section emulates one of the ideal current distributions, 'cosine-theta' or 'overlapping ellipses' [1]. However, a completely different, alternative configuration giving the same result is possible. Theoretically, the perfectly uniform transverse magnetic field across the whole aperture can be obtained with the use of special solenoidal coils, so called 'tilted coils' (Fig. 1b), where the ideal, 'cosine-theta' distribution of axial (along Z-axis) current density can be brought about in a simple way.

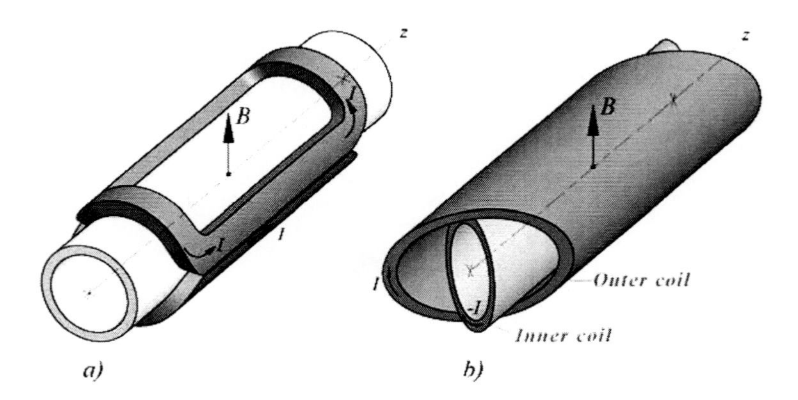

Fig. 1. Production of transverse and uniform magnetic fields by means of (***a***) curved saddle coils ('dipole') and (***b***) a pair of tilted solenoidal coils oppositely tilted and energized with opposite polarity.

Let us consider the simplest two-tilted-coil magnet, Fig. 1b. The cylindrical coils are concentric and nested, their turns are rotated an angle, 'tilt angle', with respect to the central axis, Z. Current is flowing in opposite directions in the coils tilted at opposite angles. As a result, and as shown in Fig. 2, if the coils are long enough, the longitudinal (axial, 'solenoidal') components of the magnetic field effectively cancel each other out within the aperture and the transverse, 'dipole', component of each coil adds, leaving only a purely transverse magnetic field. This resultant field turns out to be perfectly uniform (as shown below) and noticeably depends on the tilt angle, α : the smaller tilt angle is chosen (the more tilt), the higher resultant transverse field is obtained.

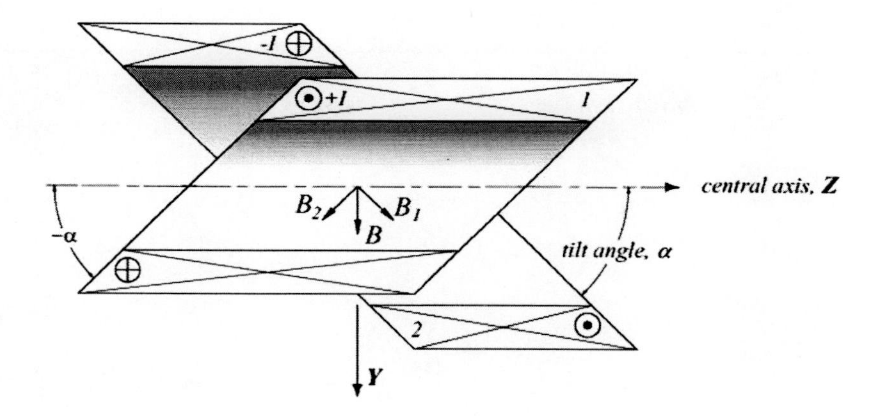

Fig. 2. Sketch of the simplest tilted coil magnet with two-coil configuration. Cross-sectional view: B_1 is the field from the inner coil, B_2 – from the outer coil, $B = B_Y$ is the resultant transverse field.

Obviously, the main disadvantage of a tilted coil magnet is that a considerable proportion of power and conductor is lost due to the axial field canceling. The advantage is that a relatively simple solenoid winding technology with a single strand superconductor can be applied. In addition, a solenoidal winding is more compact and rigidly built than that of a traditional dipole, and the cost of a tilted coil magnet manufacturing can be lower.

In the design, each tilted coil can have a single-layer winding ('helix'), as shown in Fig. 3. A multi-layer winding can also be suitable and practical, depending on the application. Both round and rectangular conductors can be applied. The turns are of elliptical shape with $1/\sin\alpha$ aspect ratio, keeping the whole coil and its aperture circular in shape. At the same time, the ellipse is not fully flat because of the helical path of the conductor, and hence each layer of the winding represents a tilted cylindrical helicoid. It is also noteworthy that there exists a technological problem of holding tilted turns in a predetermined 'tilted' position, which is evidently unstable if care is not taken – as the conductor is pulled tight during a layer-wise winding process, it has a tendency to take the shortest path: a circle lying in a plane perpendicular to the tube instead of the inclined ellipse. This problem, non-existent in the case of a conventional, non-tilted solenoid, might be a major one for a tilted coil, and it can be effectively solved in at least two ways detailed below. Let us just note here that the most obvious and 'natural' way is to hold each particular turn using a helical support structure – a tilted helix form that can be fabricated by milling a groove of the correct geometry into the tube. Such a support structure is schematically shown in Fig. 3.

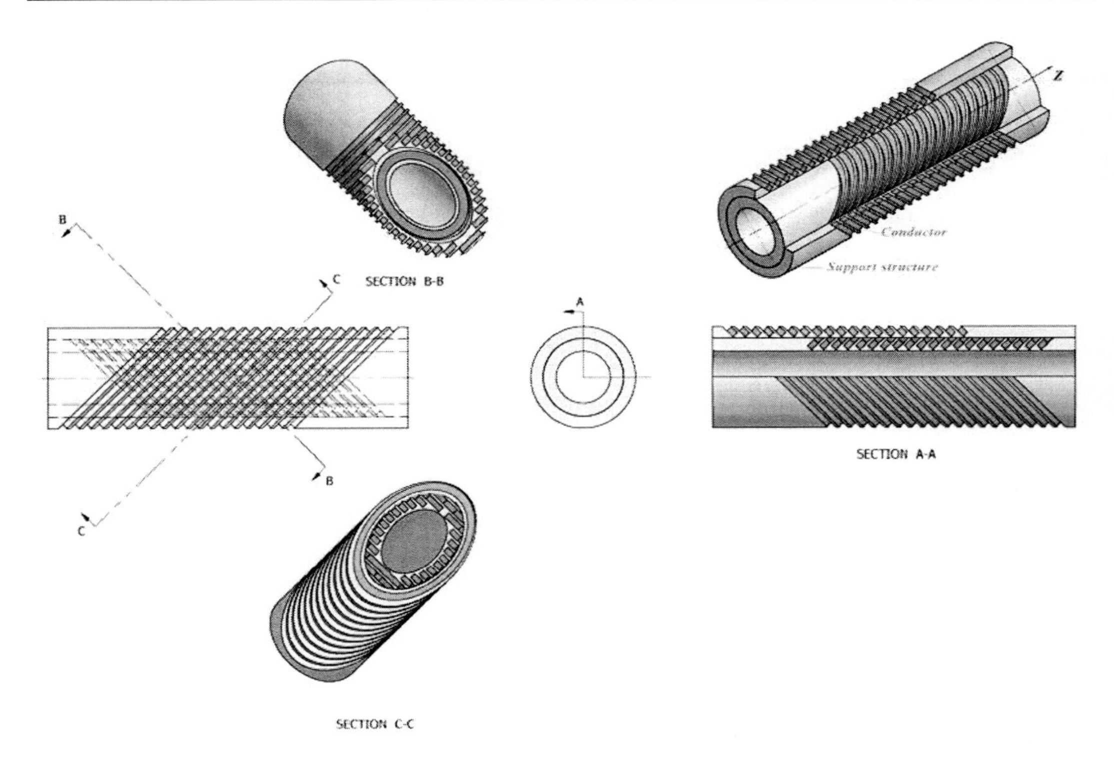

Fig. 3. 3D sketch and cross-sections of rectangular-wire-wound two-tilted-coil magnet with support structure.

In order to obtain a higher transverse field with higher uniformity using tilted coils, more pairs of such coils can be employed, Fig. 4. By grading the conductor, different current densities can be used in the coils. In doing so, the total number of coils can be either even or odd. However, in both cases, the current densities must be properly adjusted to fully cancel the axial component of the resultant magnetic field out so as to attain a purely transverse direction of the field. Obviously, by simply changing the opposite current polarity in the odd-numbered coils to the same as in their even-numbered neighbors (see Figs. 2,4), a purely axial magnetic field can be obtained instead of the transverse one. A possibility to easily change the field direction from purely transverse to purely axial in the same magnet is another real advantage of the tilted coil magnet concept. Moreover, by varying currents in the coils, any direction of the field may be attainable. This may be attractive, for instance, for rotation experiments with a large access port perpendicular to the field [2].

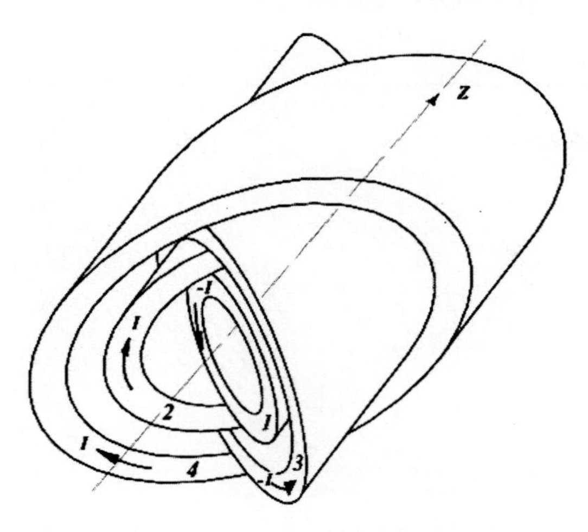

Fig. 4. 3D sketch of multiple-tilted-coil-magnet configuration. Arrows show the current direction. The coils can be wire-wound multi-layer, wire-wound one-layer ('helix') and/or Bitter type (see below).

To our knowledge, the idea to use wire-wound concentric nested tilted coils for uniform transverse magnetic field generation was first published by D.I. Meyer and R. Flasck in 1970 [3]. However, they considered circular turns and, correspondingly, elliptical aperture coils only. In the mid-1970s, Prof. V.E. Keilin tried the modified idea to use elliptical turns to built a tilted-coil 'cosine-theta' wire-wound magnet with circular aperture [4]. His team attempt to implement it was made at the Kurchatov Institute, Moscow, Russia, in the mid-1990s [4], but the development was soon discontinued. A 45 degree tilt angle only was considered appropriate. The idea may have been viewed or tried in other research centers of the former USSR. It was also tried in France in 1974, where CEA/Saclay built a nested two-tilted-coil ('double helix') magnet [5]. It was a simplistic thin winding with very moderate tilt of turns [5]. No detailed information on the development and its purpose is available [5].

The tilted coil idea development was resumed in the USA at the beginning of 2000, at the National High Magnetic Field Laboratory (NHMFL), Tallahassee, Florida [4], [6]. Advanced Magnet Lab, Inc. (AML, Inc.), Florida, USA, made attempts to develop relevant winding technology [7], [8] and to build small tilted coil prototypes and, with the NHMFL help, a small practical laboratory tilted coil non-superconducting magnet [9].

The highest DC transverse field that can be achieved with an up-to-date superconducting wire-wound technology is definitely lower than 20T [4]. To obtain uniform transverse magnetic fields noticeably higher than 20T DC within a reasonably large volume of space, Bitter technology [2], [4] is appropriate, where copper alloy sheet metal is used instead of wire (Figs. 5,6). As opposed to turns of a wire-wound tilted coil, elliptical Bitter disks can be easily stacked in any orientation, maintaining each other in a predetermined position, offering herewith a strong winding, where no tilted helix form is needed as a support. The side support structure of the innermost coil is truncated at the tilt angle (Fig. 6). When installed, it supports the thick-walled inner tube and the alignment rods in a vertical position. The coil is built up of alternating conductors and insulators using one of the standard methods. The other side support structure is mounted, and tie rods are installed instead of the alignment rods. The inner tube, side support structure and tie-rods secure the coil. The other coils are built up and attached to the innermost coil along with support elements [4].

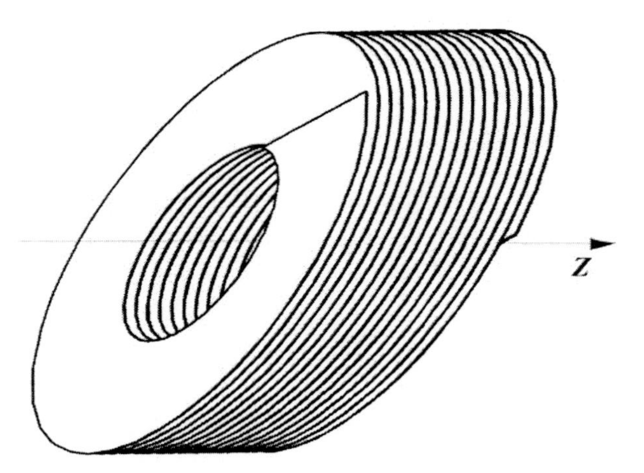

Fig. 5. Fragment of a tilted Bitter coil with helical path of metal sheet conductor. Sketch. Tie-rod holes, cooling holes and connections between elliptical Bitter disks forming the winding are not shown.

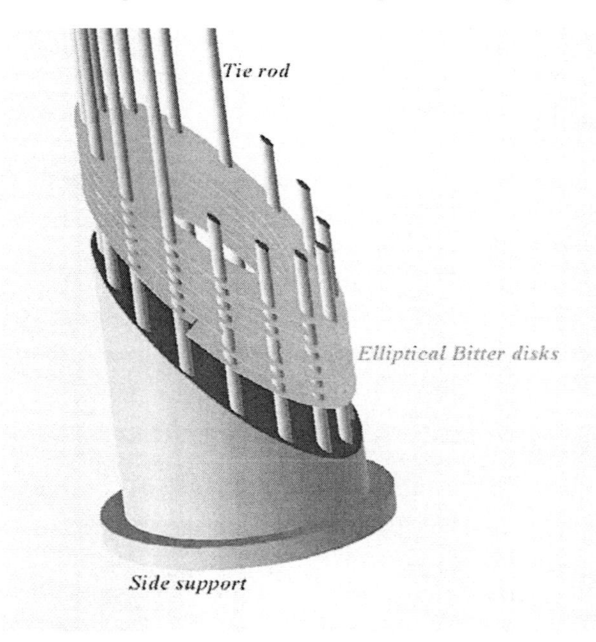

Fig. 6. Side view of a tilted Bitter coil. Fragment. The support structure, tie rods and elliptical Bitter disks with helical path of the winding are shown.

The idea to apply the Bitter technology to tilted coils was suggested by A.V. Gavrilin at the beginning of 2000 at the NHMFL, and the concept was then developed by A.V. Gavrilin and M.D. Bird during 2000-2002, when first practical designs were suggested [4] and a US patent application was submitted (the patent awarded in 2005 [10]). A mathematical method by A.M. Akhmeteli was used in this development [4].

The idea of tilted coils made with wire-wound and advanced technologies can be also applied to resistive pulse magnets providing the highest fields [6].

II Wire-wound Tilted Coil as a Generalized "Cosine-Theta" Configuration

It is rather easy to prove that wire-wound circular tilted coils have a generalized "cosine-theta" distribution of axial current density. In order to do so, it is sufficient to consider a fragment of one layer of the winding (Fig. 7). For simplicity, the wire can be considered round without loss of generality.

Let us assume that the wire is wound in such a way that centers of cross-sections of the wire lie on the following parametric curve (a tilted cylindrical helicoid, Fig. 7):

$$\mathbf{P}(\theta) = \{R\cos\theta, R\sin\theta, Rq\sin\theta + w\theta\}, \tag{1}$$

R is the average radius of the winding, q and w are constants here (see (9) below), θ is the azimuth angle in the cylindrical system of coordinates.

An increase of θ by 2π leads to a shift by a vector (independent of θ): $\mathbf{V}_1 = \{0,0,2\pi w\}$. A vector tangent to curve $\mathbf{P}(\theta)$ may be found by differentiation of (1) with respect to θ:

$$\mathbf{V}_2 = \mathbf{V}_2(\theta) = \{-R\sin\theta, R\cos\theta, Rq\cos\theta + w\}, \tag{2}$$

and the vector of current flowing along the wire can be determined:

$$\mathbf{I} = \mathbf{I}(\theta) = I\cdot\mathbf{V}_2/V_2, \tag{3}$$

where $I = |\mathbf{I}(\theta)| = const$ is nothing else but the value of the current in the wire; $V_1 = |\mathbf{V}_1| = const$ and $V_2 = |\mathbf{V}_2|$ are the magnitudes of the relevant vectors. The average current density vector in the winding equals

$$\mathbf{j} = \mathbf{j}(\theta) = \mathbf{I}/(d\delta), \tag{4}$$

where d is the diameter of the wire, and $\delta = \delta(\theta)$ is the distance between tangents to curve $\mathbf{P}(\theta)$ drawn in two points where angle θ differs by 2π (Fig. 7).

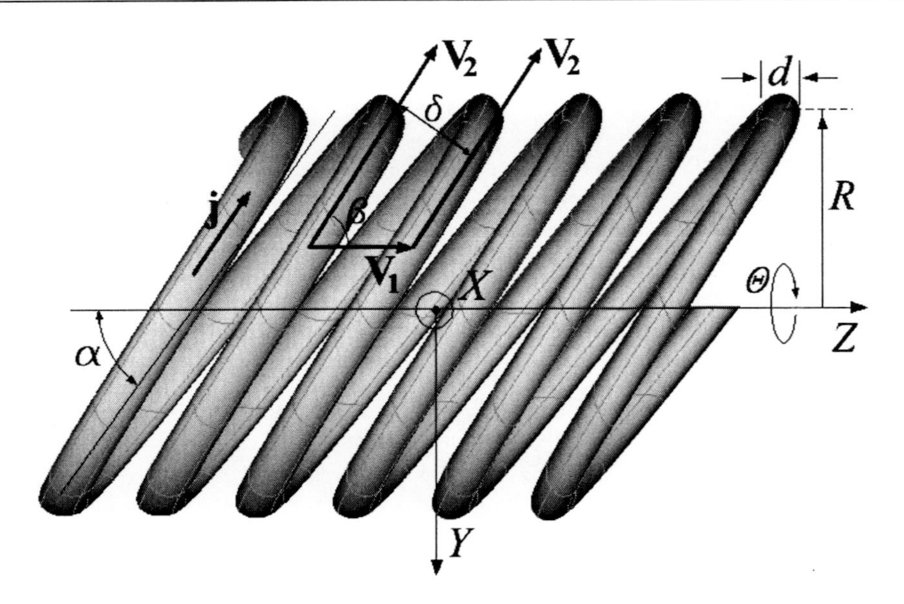

Fig. 7. Fragment of one layer of a tilted coil wound with a circular cross-section conductor (wire).

It should be remembered that the magnitude of the average current density, $j = |\mathbf{j}|$, depends on θ, although neither the current, I, nor the wire cross-section depends on θ. This is due to the fact that the winding cannot be uniformly tight in this geometry, and a small gap between two adjacent turns of the winding is bound to take place. This gap, g, depends on θ:

$$g(\theta) \approx \delta(\theta) - d \quad (5)$$

The gap is minimal at $\theta \approx k\pi$, where k is an integer; in particular, in the vicinity of these points the adjacent turns can touch each other if the winding is closely packed. The gap is widest at $\theta = \pi/2 + k\pi$, and it noticeably depends on the tilt angle α (Fig. 7) – the smaller the tilt angle, the greater the gap.

Evidently,

$$\delta = V_1 \sin\beta = |\mathbf{V}_1 \times \mathbf{V}_2| / V_2 = 2\pi w R / V_2 , \quad (6)$$

where $\beta = \beta(\theta)$ is the angle between vectors $\mathbf{V}_1$ and $\mathbf{V}_2$ at a given value of θ (see Fig. 6).

Consequently, the average current density vector can be expressed as

$$\mathbf{j} = I \cdot \mathbf{V}_2 / (V_2 d \delta) = I \cdot \mathbf{V}_2 / (2\pi w R d) , \quad (7)$$

and thus the axial component of the vector turns out to be

$$j_Z = I\frac{Rq\cos\theta + w}{2\pi wRd} = j_{Z\,\max}\cdot\cos\theta + j_Z^{const} \qquad (8)$$

It is noteworthy, albeit almost obvious, that the constant (independent of tilt angle) component of the axial current arises due to the helical path of wire – it vanishes if each turn is considered a flat ellipse. Technically, the presence of the constant component of axial current is a distinction, compared to classic cosine-theta coils, which basically are not supposed to have such a "makeweight". Nevertheless, this is "a distinction without a difference", which does not deteriorate the ideal uniformity of transverse field within the bore. It is evident that this constant axial component of the current is present in ordinary coils without tilt.

It is easy to see how the parameters of curve $\mathbf{P}^{(\theta)}$ are related to the parameters of the one-layer winding: the external and the internal radii equal $R+d/2$ and $R-d/2$, respectively; $q=\cot\alpha$. Parameter w may be found from the following equation:

$$d \approx \delta(0) = \min(\delta(\theta)) = \min\left(\frac{2\pi wR}{V_2}\right) = $$
$$= \min\left(\frac{2\pi wR}{\sqrt{R^2+(Rq\cos(\theta)+w)^2}}\right) = \frac{2\pi wR}{\sqrt{R^2+(Rq+w)^2}}, \qquad (9)$$

assuming that the coil is wound as tightly as possible.

As already mentioned, the wire cross-section does not have to be circular, it can be rectangular or of any other reasonable shape. Furthermore, theoretically, even a wire of variable cross-section may be used. In both cases the field uniformity is not affected, as the current is still constant along the wire, and the above proof holds. These options may have some advantages. For example, the gaps (see (5)) can be eliminated using a wire of a properly varying section. As a result, the total resistance of the winding can be decreased in resistive coils, and the mechanical strength can be increased.

Obviously, the more layers of winding are used, the higher transverse component of magnetic field can be obtained at the same current. As to the axial component of the magnetic field, also generated by a tilted coil, it can be easily cancelled out through the use of two nested oppositely tilted coils energized with opposite polarity (Figs. 1b, 2-4).

It can be easily shown that the transverse field dependence on the tilt angle is non-linear – neglecting end effects, the field decreases directly with the cosine of the tilt angle α for wire-wound tilted coils. The same dependency is approximately valid for Bitter tilted coils. Thus, the use of very small tilt angles, 30° and less, is not so beneficial as it might look, especially if we also take into account that the decrease of tilt angle carries a penalty: probably more difficult assembly, increased end effects and, maybe, lower strength.

Using (1), one can numerically calculate the magnetic field generated by a tilted coil magnet. In doing so, we can investigate how minor natural errors in the conductor positioning within a tilted winding can affect the field uniformity in cases where a very high and special quality of the field is required, e.g., in magnets for modern accelerators. This problem has not been investigated in full yet. At the same time, there are no clear reasons to believe that it is more complex than that for traditional curved-saddle dipoles.

III Wire-Wound Tilted Coil Magnet Design Example

A good example of a high field superconducting tilted coil magnet is cited in work [6]. This is a four-coil magnet; its characteristics are given in Table I. A conventional NbTi/Cu composite superconducting wire of 1.065 mm diameter is used for the tilted multi-layer windings. In the magnet, the innermost and outermost coils (1&4) differ from the inner coils (2&3) in the total number of winding layers only; the inner coils (2&3) have roughly twice as many layers as 1&4, as this maximizes the length of the zone where the field is uniform. It is noteworthy that the influence of the length of the four tilted coil magnet on its field characteristics is practically excluded if the magnet is sufficiently long (> 1.5 m). The magnet can provide uniform transverse magnetic field of over 8T within a 56 mm cold bore. As can be inferred from Fig. 8, the field uniformity zone is about 400 mm long. With lesser tilt angle (~30°), up-to-date Nb_3SN superconductors [11] and additional design optimization, 15-16T transverse field can be possible within the same aperture.

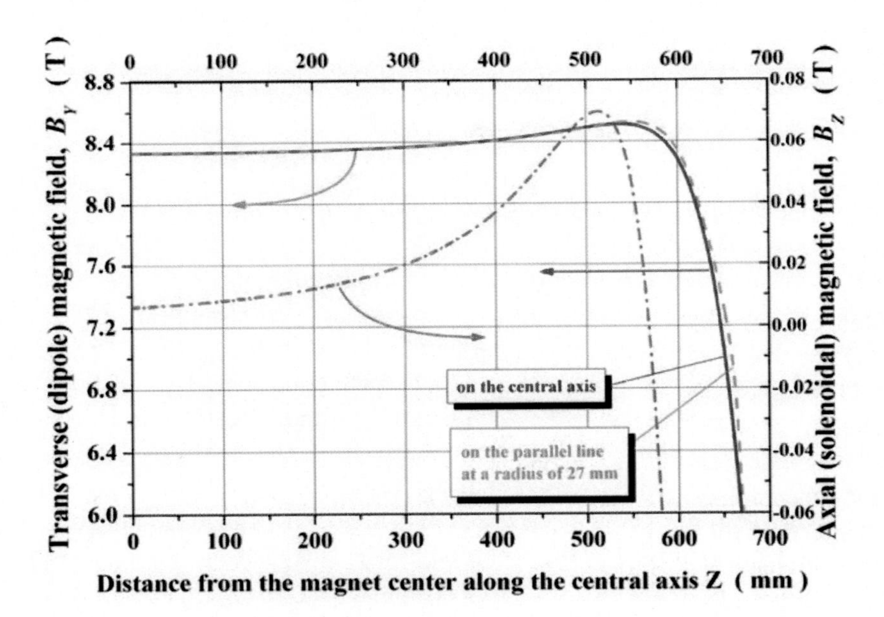

Fig. 8. Axial distribution of transverse, B_Y, and axial, B_Z, magnetic fields of the four tilted coil superconducting magnet. Transverse field is given on the central axis (X=0, Y=0, solid line) and on the parallel line (X=0, Y=27 mm, dashed) in the vicinity of the magnet inner surface at a radius of 27 mm. Axial (undesired) field (dash-dot) is given on the parallel line only; it is even lower on the central axis.

Generally speaking, the field uniformity that can be obtained with tilted coils would be certainly sufficient for many experiments in solid state physics, neutron scattering or for material tests. At the same time, as mentioned above, it will take further research to decide if use of tilted coil magnets in accelerators is practical.

It is also noteworthy that there are no clear reasons to think that quench behavior of superconducting tilted coils will differ much from that of conventional superconducting magnets using similar conductors. So, superconducting tilted coil magnets can be quench protected reliably by the traditional methods presently used in large scale application of superconductivity [12].

Table I. Characteristics of four-tilted-coil superconducting magnet [6].

Total number of coils	4
Tilt angle (degrees)	45 {-45,+45,-45,+45}
Gap between coils	None
Conductor, round, single strand	1.065 mm NbTi/Cu
Cu/SC ratio	1.65
Insulated conductor diam (mm)	1.4
Critical current of the conductor	~490A @ 10T & 1.9K ~350 A @ 8.3T & 4.2K
Inner radius (mm)	28
Outer radius (mm)	115
Length (mm)	1614
Total number of winding layers	62 {11+20+20+11}
Total number of turns in each layer	700
Rated current (A) / polarity in coils	421 {+421,-421,+421,-421}
Central field @ 421A (T)	8.332

IV Resistive Bitter Tilted Coils

Bitter tilted coil magnets enable generation of considerably higher persistent transverse magnetic fields than what superconducting dipole magnets can produce in the same volume of space. This makes them attractive for a number of physical experiments, in particular, for testing lengthy samples of new high field and high current density superconductors in transverse magnetic fields to 16-20T and even higher. However, Bitter tilted coils (Figs. 5,6) are much more complicated for analysis than wire-wound tilted coils.

The magnetic field from a Bitter tilted coil can be adequately approximated by the field of a number of identical elliptical disks parallel to each other – the same approach is used for conventional Bitter solenoids [13]. In our analysis, we neglect the presence of numerous holes and slits in the disks. Let us calculate the current density distribution in one elliptical Bitter disk. While for a conventional circular Bitter disk the current density distribution obeys a very simple analytical formula [13], this is not the case for the current density in an elliptical disk (Fig. 9) of a tilted Bitter coil (Figs. 5,6). However, a more complex exact solution was found for the latter for an arbitrary tilt angle.

Let us consider an elliptical disk defined by the following equations in Cartesian coordinates x_1, x_2, x_3:

$$p_1^2 \le wx_1^2 + x_2^2 \le p_2^2,$$
$$-\frac{d}{2} \le x_3 \le \frac{d}{2}. \tag{10}$$

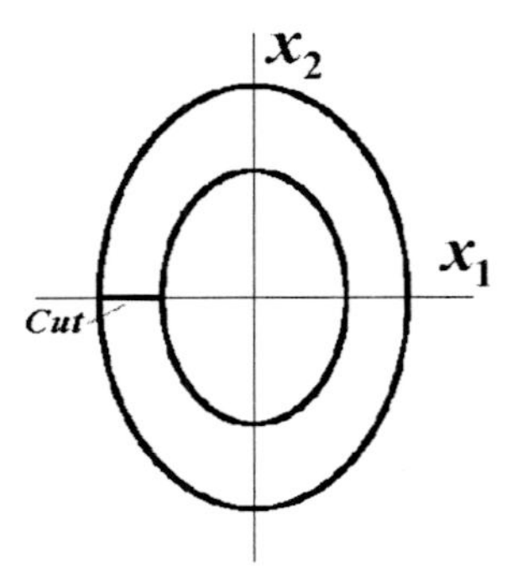

Fig. 9. Sketch of a Bitter disk bounded with homothetic ellipses with a coordinate system aligned with it.

Here $w > 1$. It is noteworthy that due to the circular shape of the coil and its aperture the outer and inner ellipses of the disk are homothetic.

We assume that the disk is cut in the half plane $x_1 < 0$, and static electric potentials Φ_1 and Φ_2 are applied to the surfaces of the cut. Due to the symmetry of the problem, the potential (and, consequently, the current density) does not depend on x_3 (we assume that the conductivity σ is constant within the disk). As the current density $\mathbf{j}(x_1, x_2)$ is static, the continuity equation has the following form:

$$\nabla \mathbf{j}(x_1, x_2) = 0. \tag{11}$$

As

$$-\nabla\Phi(x_1, x_2) = \mathbf{E}(x_1, x_2) = \mathbf{j}(x_1, x_2)/\sigma, \tag{12}$$

where $\mathbf{E}(x_1, x_2)$ is the electric field strength, the problem may be stated as follows.

We seek the partial solution $\Phi(x_1, x_2)$ of the equation $\Delta\Phi = 0$ in the area limited by two homothetic ellipses that are defined by the equations:

$$wx_1^2 + x_2^2 = p_1^2,$$
$$wx_1^2 + x_2^2 = p_2^2, \tag{13}$$

with a cut along the abscissa axis (in the left half-plane).

On the upper and the lower banks of the cut, the function equals π and $-\pi$, respectively; these values were chosen for convenience; if necessary, this solution may be modified in an obvious way to accommodate for the actual values of Φ_1 and Φ_2. At the elliptic boundaries, the projection of the gradient of this function on the normal to the ellipses equals zero (as there is no charge flow through these boundaries).

Using a standard procedure (see, e.g., [14]), one can prove that the function sought realizes an extremum of the functional

$$J(\Phi) = \int_S dS(\nabla\Phi)^2, \tag{14}$$

where S is the area between the ellipses, on the set of functions that have the same values on the banks of the cut ($\pm\pi$).

The equation $\Delta\Phi = 0$ may be solved by the separation of variables in non-orthogonal coordinates (cf. [15]):

$$u = \ln r + \frac{1}{2}\ln((w-1)\cos(2\varphi) + w + 1), \quad v = \varphi, \tag{15}$$

where r and φ are the polar coordinates. This method is equivalent to solving the equation by the Ritz method (see, e.g., [14]) with the following functions (multipoles) used as the basis:

$$(\alpha \ln r + \beta)(\gamma\varphi + \delta), \tag{16}$$

$$Cr^{\pm n} \exp(\pm in\varphi), \tag{17}$$

where $\alpha, \beta, \gamma, \delta, C$ are constants. The sought-for solution Φ will be a linear combination of these solutions. In view of the boundary conditions on the banks of the cut, the solution may be sought in the form:

$$\Psi(r,\varphi) = \varphi + \sum_{n\neq 0} c_n r^n \sin(n\varphi), \tag{18}$$

where n is an integer (possibly, negative).

The coefficients c_n are evaluated from the condition of extremum of the functional. In polar coordinates, the gradient has the following form:

$$\nabla\Psi = \frac{\partial\Psi}{\partial r}\mathbf{i}_r + \frac{1}{r}\frac{\partial\Psi}{\partial\varphi}\mathbf{i}_\varphi , \tag{19}$$

$$\nabla\varphi = \frac{1}{r}\mathbf{i}_\varphi , \tag{20}$$

$$\nabla(r^n \sin(n\varphi)) = nr^{n-1}\sin(n\varphi)\mathbf{i}_r + r^{n-1}n\cos(n\varphi)\mathbf{i}_\varphi , \tag{21}$$

$$\int dS(\nabla\Phi)^2 = \int dS(\nabla(\varphi + \sum_{n\neq 0} c_n r^n \sin(n\varphi)))^2. \tag{22}$$

We may write the following:

$$\nabla\Psi = \sum_n g_n r^{n-1}(\sin(n\varphi)\mathbf{i}_r + \cos(n\varphi)\mathbf{i}_\varphi), \tag{23}$$

where

$$g_n = \begin{cases} 1 & n = 0 \\ nc_n & n \neq 0 \end{cases} . \tag{24}$$

Then

$$\begin{aligned}(\nabla\Psi)^2 &= \sum_{m,n} g_n g_m r^{m+n-2}(\sin(n\varphi)\sin(m\varphi) + \cos(n\varphi)\cos(m\varphi)) = \\ &= \sum_{m,n} g_n g_m r^{m+n-2}\cos((m-n)\varphi)\end{aligned} , \tag{25}$$

$$\int_S (\nabla\Psi)^2 dS = \sum_{m,n} g_n g_m A_{nm} , \tag{26}$$

$$\begin{aligned} A_{nm} &= \int_S r^{m+n-2}\cos((m-n)\varphi)dS = \int_{-\pi}^{\pi} d\varphi\cos((m-n)\varphi)\int_{r_1(\varphi)}^{r_2(\varphi)} rdr\cdot r^{m+n-2} = \\ &= \begin{cases} \int_{-\pi}^{\pi} d\varphi\cos((m-n)\varphi)(\ln r_2(\varphi) - \ln r_1(\varphi)) & (m+n=0) \\ \int_{-\pi}^{\pi} d\varphi\cos((m-n)\varphi)\frac{r_2^{m+n}(\varphi) - r_1^{m+n}(\varphi)}{m+n} & (m+n\neq 0) \end{cases} \end{aligned} , \tag{27}$$

where $r_2(\varphi)$ and $r_1(\varphi)$ are the coordinates r of the points of the larger and the lesser ellipses with coordinate φ, correspondingly.

The equation of the larger ellipse is:

$$wx_1^2 + x_2^2 = p_2^2$$

or

$$wr^2 \cos^2 \varphi + r^2 \sin^2 \varphi = p_2^2, \tag{28}$$

$$r_2^2(\varphi) = \frac{p_2^2}{w\cos^2 \varphi + \sin^2 \varphi} = \frac{p_2^2}{1+(w-1)\cos^2 \varphi} = \frac{p_2^2}{1+(w-1)\dfrac{\cos(2\varphi)+1}{2}} =$$
$$= \frac{2p_2^2}{(w-1)\cos(2\varphi)+w+1} \tag{29}$$

Similarly,

$$r_1^2(\varphi) = \frac{2p_1^2}{(w-1)\cos(2\varphi)+w+1}. \tag{30}$$

Thus, if $m+n=0$, then

$$A_{nm} = \int_{-\pi}^{\pi} d\varphi \cos((m-n)\varphi) \cdot \frac{1}{2}\left(\ln\frac{2p_2^2}{(w-1)\cos(2\varphi)+w+1} - \ln\frac{2p_1^2}{(w-1)\cos(2\varphi)+w+1} \right) =$$
$$= \int_{-\pi}^{\pi} d\varphi \cos((m-n)\varphi) \cdot \ln\frac{p_2}{p_1} = \delta_{mn} \cdot 2\pi \cdot \ln\frac{p_2}{p_1}, \tag{31}$$

where δ_{mn} is the Kronecker's delta.

If $m+n \neq 0$, then

$$A_{nm} = \int_{-\pi}^{\pi} d\varphi \cos((m-n)\varphi) \cdot \frac{1}{m+n}\left(\left(\frac{2p_2^2}{(w-1)\cos(2\varphi)+w+1}\right)^{\frac{m+n}{2}} - \left(\frac{2p_1^2}{(w-1)\cos(2\varphi)+w+1}\right)^{\frac{m+n}{2}} \right) =$$
$$= \frac{2^{\frac{m+n}{2}}}{m+n}(p_2^{m+n} - p_1^{m+n}) \int_{-\pi}^{\pi} d\varphi \cos((m-n)\varphi)((w-1)\cos(2\varphi)+w+1)^{-\frac{m+n}{2}}. \tag{32}$$

If we only leave harmonics with numbers from $-k$ to k, then, differentiating quadratic form $\sum_{n,m=-k}^{k} A_{nm} g_n g_m$ with respect to g_n ($n \neq 0$) and equating the results to zero, we obtain a system of linear equations that determines g_n ($n \neq 0$).

Let us consider the integral

$$\int_{-\pi}^{\pi} d\varphi \cos((m-n)\varphi)((w-1)\cos(2\varphi)+w+1)^{-\frac{m+n}{2}} . \tag{33}$$

If $(m-n)$ is odd, this integral equals zero: the function $((w-1)\cos(2\varphi)+w+1)^{-\frac{m+n}{2}}$ is periodic with period π, so there are even harmonics only ($\exp(\pm 2ik\varphi)$, where k is integer) in its Fourier series, and the integral equals the coefficient of an odd harmonic to a factor. Therefore, in the series for Φ we may leave even terms only:

$$\Phi(r,\varphi) = \varphi + \sum_{k \neq 0} c_{2k} r^{2k} \sin(2k\varphi) , \tag{34}$$

where k is integer.

Let $m = 2l$, $n = 2k$; the integral has the following form:

$$\int_{-\pi}^{\pi} d\varphi \cos(2(l-k)\varphi)((w-1)\cos(2\varphi)+w+1)^{-l-k} = \frac{1}{2}\int_{-2\pi}^{2\pi} du \cos((l-k)u)((w-1)\cos(u)+w+1)^{-l-k} =$$

$$= \int_{-\pi}^{\pi} du \cos((l-k)u)((w-1)\cos(u)+w+1)^{-l-k} . \tag{35}$$

So, let us evaluate

$$I = \int_{-\pi}^{\pi} du \cos((l-k)u)(\cos(u)+q)^{-l-k} = \mathrm{Re} \int_{-\pi}^{\pi} du \exp(i(l-k)u)(\cos(u)+q)^{-l-k}, \tag{36}$$

where

$$q = \frac{w+1}{w-1} > 0 . \tag{37}$$

Let $z = \rho \exp(iu)$. On the unit circle,

$$dz = \rho i \exp(iu) du = i \exp(iu) du . \tag{38}$$

$$I_1 = \int_{-\pi}^{\pi} du \exp(i(l-k)u)(\cos(u)+q)^{-l-k} = \oint \frac{dz}{i\exp(iu)} \cdot \exp(i(l-k)u)\cdot(\cos(u)+q)^{-l-k} =$$

$$= \oint dz(-i)z^{l-k-1}\left(\frac{z+z^{-1}}{2}+q\right)^{-l-k} = -i\cdot 2^{l+k}\oint dz\cdot z^{2l-1}(z^2+2qz+1)^{-l-k}. \qquad (39)$$

The integration contour is the unit circle. Inside the unit circle, there may be poles in the points $z=0$ and $z=-q+\sqrt{q^2-1}$.

Let us first consider the case where $l+k<0$. Let us evaluate the integral

$$\oint dz\cdot z^m (z^2+2qz+1)^n, \qquad (40)$$

where $n>0$. If at the same time $m\geq 0$, there are no singular points inside the unit circle, so the integral equals zero. Let us consider the integral

$$\oint dz\cdot z^{-m}(z^2+2qz+1)^n, \qquad (41)$$

where $n,m>0$.

$$(z^2+2qz+1)^n = ((z+q)^2-(q^2-1))^n = \sum_{j=0}^{n}\binom{n}{j}(z+q)^{2j}(-(q^2-1))^{n-j}, \qquad (42)$$

where $\binom{n}{j}=\frac{n!}{j!(n-j)!}$ is the number of all combinations of n elements taken j at a time.

In this connection, let us evaluate the integral

$$\oint dz\cdot z^{-m}(z+q)^k, \qquad (43)$$

where $m>0$, $k\geq 0$. It has only one pole of order m in the zero point. If inside the integration contour the only singular points are poles, integral $\oint_C dz\cdot f(z)$ equals $2\pi i\sum_k \text{Res}\, f(z_k)$, where z_k are poles, and $\text{Res}\, f(z_k)$ is the residue in the pole z_k. A residue in the point a that is a pole of order m equals

$$\text{Res}\, f(a) = \frac{1}{(m-1)!}\lim_{z\to a}\frac{d^{m-1}}{dz^{m-1}}((z-a)^m f(z)). \qquad (44)$$

In our case the residue in zero equals

$$\frac{1}{(m-1)!}\lim_{z\to 0}\frac{d^{m-1}}{dz^{m-1}}(z+q)^k \quad . \tag{45}$$

If $k<m-1$, the residue equals zero, and if $k\ge m-1$, it equals

$$\frac{1}{(m-1)!}k(k-1)\ldots(k-m+2)\cdot q^{k-m+1}=\frac{1}{(m-1)!}\frac{k!}{(k-m+1)!}\cdot q^{k-m+1}=\binom{k}{m-1}\cdot q^{k-m+1} \quad . \tag{46}$$

Thus, the evaluation of the integral $\oint dz\cdot z^m(z^2+2qz+1)^n$, where $n>0$, is completed. Now let us consider the integral

$$\oint dz\cdot z^m(z^2+2qz+1)^{-n} \quad , \tag{47}$$

where $n>0$.

$$\begin{aligned}z^2+2qz+1&=\left(z+q+\sqrt{q^2-1}\right)\left(z+q-\sqrt{q^2-1}\right)=-\left(z+q+\sqrt{q^2-1}\right)\left(\sqrt{q^2-1}-z-q\right)=\\&=-2\sqrt{q^2-1}\left(\frac{1}{2}+\frac{z+q}{2\sqrt{q^2-1}}\right)\cdot 2\sqrt{q^2-1}\left(\frac{1}{2}-\frac{z+q}{2\sqrt{q^2-1}}\right) \quad .\end{aligned} \tag{48}$$

Then

$$\frac{1}{z^2+2qz+1}=\frac{-1}{4(q^2-1)\left(\frac{1}{2}+\frac{z+q}{2\sqrt{q^2-1}}\right)\left(\frac{1}{2}-\frac{z+q}{2\sqrt{q^2-1}}\right)}=-\frac{1}{4(q^2-1)}ab \quad , \tag{49}$$

where

$$a=\frac{1}{\frac{1}{2}+\frac{z+q}{2\sqrt{q^2-1}}}\ , \quad b=\frac{1}{\frac{1}{2}-\frac{z+q}{2\sqrt{q^2-1}}} \quad . \tag{50}$$

Note that $a+b=ab$.

Let us prove that if $a+b=ab$ and $n>0$, then

$$(ab)^n=\sum_{i=1}^{n}c_i^{[n]}(a^i+b^i) \quad , \tag{51}$$

where $c_i^{[n]}$ is defined by the following recurrence formulae:

$$c_1^{[1]} = 1,$$

$$c_i^{[n]} = (1+\delta_{i1}) \sum_{j=i-1+\delta_{i1}}^{n-1} c_j^{[n-1]} \quad \text{for } n>1 \text{ (}\delta_{ij} \text{ is the Kronecker's delta).} \tag{52}$$

Proof.

It is evident that the statement is true for $n=1$. Let us assume that the statement is true for $n-1$ ($n>1$) and prove that it is true for n.

As $a+b=ab$, we obtain $b=\dfrac{a}{a-1}$, $a=\dfrac{b}{b-1}$.

Then

$$\begin{aligned}
(a+b)(a^i+b^i) &= a^{i+1}+ba^i+ab^i+b^{i+1} = a^{i+1}+\frac{a}{a-1}a^i+\frac{b}{b-1}b^i+b^{i+1} = \\
&= a^{i+1}+\frac{a^{i+1}}{a-1}+\frac{b^{i+1}}{b-1}+b^{i+1} = \frac{a^{i+2}}{a-1}+\frac{b^{i+2}}{b-1} = \frac{a^{i+2}-a+a}{a-1}+\frac{b^{i+2}-b+b}{b-1} = \\
&= a(1+a+\ldots+a^i)+b+b(1+b+\ldots+b^i)+a =
\end{aligned} \tag{53}$$

$$= (a+a^2+\ldots+a^{i+1})+b+(b+b^2+\ldots+b^{i+1})+a = \sum_{k=1}^{i+1}(a^k+b^k)+a+b.$$

$$\begin{aligned}
a^n b^n &= \sum_{i=1}^{n} c_i^{[n]}(a^i+b^i) = (a+b)\sum_{i=1}^{n-1} c_i^{[n-1]}(a^i+b^i) = \\
&= \sum_{i=1}^{n-1} c_i^{[n-1]}\left(\sum_{k=1}^{i+1}(a^k+b^k)+a+b\right) = \text{(we change the order of summation)} = \\
&= \left(\sum_{k=1}^{n}(a^k+b^k) \sum_{i=\max(1,k-1)}^{n-1} c_i^{[n-1]}\right) + \sum_{i=1}^{n-1} c_i^{[n-1]}(a+b).
\end{aligned} \tag{54}$$

Hence,

$$c_i^{[n]} = \sum_{j=\max(1,i-1)}^{n-1} c_j^{[n-1]} + \delta_{i1}\sum_{j=1}^{n-1} c_j^{[n-1]} \quad (n>1),$$

or

$$c_i^{[n]} = (1+\delta_{i1}) \sum_{j=\max(1,i-1)}^{n-1} c_j^{[n-1]} = (1+\delta_{i1}) \sum_{j=i-1+\delta_{i1}}^{n-1} c_j^{[n-1]}. \tag{55}$$

This completes the proof of the statement.

Thus, for $n>0$

$$\oint dz\cdot z^m(z^2+2qz+1)^{-n}=$$

$$=\oint dz\cdot z^m\left(-\frac{1}{4(q^2-1)}\right)^n\frac{1}{\left(\frac{1}{2}+\frac{z+q}{2\sqrt{q^2-1}}\right)^n}\frac{1}{\left(\frac{1}{2}-\frac{z+q}{2\sqrt{q^2-1}}\right)^n}=$$

$$=\oint dz\cdot z^m\left(-\frac{1}{4(q^2-1)}\right)^n\sum_{i=1}^n c_i^{[n]}\left(\left(\frac{1}{\frac{1}{2}+\frac{z+q}{2\sqrt{q^2-1}}}\right)^i+\left(\frac{1}{\frac{1}{2}-\frac{z+q}{2\sqrt{q^2-1}}}\right)^i\right)=$$

$$=\left(-\frac{1}{4(q^2-1)}\right)^n\sum_{i=1}^n c_i^{[n]}\left(2\sqrt{q^2-1}\right)^{(i)}\oint dz\cdot z^m\left(\left(\frac{1}{z+q+\sqrt{q^2-1}}\right)^i+(-1)^i\left(\frac{1}{z+q-\sqrt{q^2-1}}\right)^i\right). \tag{56}$$

Therefore, it is sufficient to evaluate the integrals

$$\oint dz\cdot z^m\frac{1}{\left(z+q\pm\sqrt{q^2-1}\right)^n}\quad (n>0). \tag{57}$$

Let us first evaluate the integral

$$\oint dz\cdot z^m\frac{1}{\left(z+q+\sqrt{q^2-1}\right)^n}. \tag{58}$$

It may only have a pole in zero. The relevant residue equals zero if $m\geq 0$. Let us evaluate the residue in zero for the integral $\oint dz\cdot z^{-m}\frac{1}{\left(z+q+\sqrt{q^2-1}\right)^n}$ ($m>0$), it equals

$$\frac{1}{(m-1)!}\lim_{z\to 0}\frac{d^{m-1}}{dz^{m-1}}\left(z+q+\sqrt{q^2-1}\right)^{-n}=\frac{1}{(m-1)!}(-n)(-n-1)\ldots(-n-m+2)\left(q+\sqrt{q^2-1}\right)^{-n-m+1}=$$

$$=\frac{1}{(m-1)!}(-1)^{m-1}\frac{(n+m-2)!}{(n-1)!}\left(q+\sqrt{q^2-1}\right)^{-n-m+1}. \tag{59}$$

Now let us evaluate the integral

$$\oint dz \cdot z^m \frac{1}{\left(z+q-\sqrt{q^2-1}\right)^n} \quad (n>0). \tag{60}$$

It may have poles in zero and in the point $z=-q+\sqrt{q^2-1}$. The residue in zero is evaluated as above, i.e., it equals zero if $m \geq 0$, and the residue in zero for the integral $\oint dz \cdot z^{-m} \frac{1}{\left(z+q-\sqrt{q^2-1}\right)^n}$ ($m>0$) equals

$$\frac{1}{(m-1)!}\lim_{z\to 0}\frac{d^{m-1}}{dz^{m-1}}\left(z+q-\sqrt{q^2-1}\right)^{-n} = \frac{1}{(m-1)!}(-n)(-n-1)\ldots(-n-m+2)\left(q-\sqrt{q^2-1}\right)^{-n-m+1} =$$
$$= \frac{1}{(m-1)!}(-1)^{m-1}\frac{(n+m-2)!}{(n-1)!}\left(q-\sqrt{q^2-1}\right)^{-n-m+1}. \tag{61}$$

Now let us evaluate the residue of the integral $\oint dz \cdot z^m \frac{1}{\left(z+q-\sqrt{q^2-1}\right)^n}$ in the point $z=-q+\sqrt{q^2-1}$. It equals

$$\frac{1}{(n-1)!}\lim_{z\to -q+\sqrt{q^2-1}}\frac{d^{n-1}}{dz^{n-1}}z^m = \frac{1}{(n-1)!}(m)(m-1)\ldots(m-n+2)\left(-q+\sqrt{q^2-1}\right)^{m-n+1} =$$
$$= \begin{cases} \dfrac{1}{(n-1)!}\dfrac{m!}{(m-n+1)!}\left(-q+\sqrt{q^2-1}\right)^{m-n+1} & (m \geq n-1) \\ 0 & (0 \leq m < n\text{-}1) \\ \dfrac{1}{(\text{n-1})!}m(m-1)\ldots(m-n+2)\left(-q+\sqrt{q^2-1}\right)^{m-n+1} & (m<0) \end{cases} \tag{62}$$

Thus, a complete analytical algorithm of evaluation of the integral $\oint dz \cdot z^{2l-1}(z^2+2qz+1)^{-l-k}$ may be developed.

The amplitudes of the harmonics may also be evaluated by minimization of a different quadratic form. The modified version of the algorithm is of significant interest as it provides a simple and natural estimate of the error for a finite number of harmonics.

Let us evaluate a weighted line integral of the squared normal component of the potential gradient over the length of both boundary ellipses:

$$I_{\partial S} = \int_{\partial S}(\nabla\Psi)_n^2\sqrt{\frac{1+2q\cos(2\varphi)+q^2}{\cos(2\varphi)+q}}dl. \tag{63}$$

It is evident that the weight varies between two positive values, as $q>1$, so the integral gives an estimate of the error of the solution (the integral is zero for the exact solution, as there is no normal component of the current at the elliptical boundary). If

$$\nabla\Psi=\sum_{k=-l}^{l} g_{2k} r^{2k-1}(\sin(2k\varphi)\mathbf{i}_r+\cos(2k\varphi)\mathbf{i}_\varphi), \tag{64}$$

one can show that the normal component of the potential gradient on the boundary ellipses equals

$$(\nabla\Psi)_n=\sum_{k=-l}^{l} g_{2k} r^{2k-1}(q\sin(2k\varphi)+\sin(2(k-1)\varphi))\frac{1}{\sqrt{1+2q\cos(2\varphi)+q^2}}, \tag{65}$$

and

$$ds=\frac{\sqrt{1+2q\cos(2\varphi)+q^2}}{\cos(2\varphi)+q} r|d\varphi|. \tag{66}$$

Then one may show that

$$I_{\partial S}=\sum_{k,j=-l}^{l} g_{2k} g_{2j} D_{kj}, \tag{67}$$

where

$$\begin{aligned}D_{kj}&=\left(\frac{2}{w-1}\right)^{k+j-1/2}\left(p_2^{2k+2j-1}+p_1^{2k+2j-1}\right)\int_{-\pi}^{\pi}(\cos(2k\varphi)+q)^{-k-j-1}\cdot\\&\cdot(q\sin(2k\varphi)+\sin(2(k-1)\varphi))\cdot(q\sin(2j\varphi)+\sin(2(j-1)\varphi))d\varphi=\\&=\left(\frac{2}{w-1}\right)^{k+j-1/2}\left(p_2^{2k+2j-1}+p_1^{2k+2j-1}\right)I_{kj}.\end{aligned} \tag{68}$$

Then one can obtain:

$$\begin{aligned}I_{kj}&=\int_{-\pi}^{\pi}(\cos(u)+q)^{-k-j-1}(q\sin(ku)+\sin((k-1)u))\cdot(q\sin(ju)+\sin((j-1)u))du=\\&=\frac{1}{2}((q^2+1)I(q,k-j,-k-j-1)-q^2 I(q,k+j,-k-j-1)+qI(q,k-j-1,-k-j-1)-\\&-2qI(q,k+j-1,-k-j-1)+qI(q,k-j+1,-k-j-1)-I(q,k+j-2,-k-j-1)),\end{aligned} \tag{69}$$

where

$$I(q,m,n)=\int_{-\pi}^{\pi}\cos(mu)(\cos(u)+q)^{n}\,du. \tag{70}$$

The algorithm of evaluation of $I(q,m,n)$ is outlined above (we should just add that if $n=0$, this integral is trivial and equals $2\pi\delta_{0m}$).

It should be noted that this modified version of the algorithm has another advantage: the matrix of the relevant system of linear equations is symmetric and positive definite, so the system may be solved using a faster and numerically more stable method (Cholesky decomposition). This is important, as for ellipses with higher eccentricity, numerical stability becomes a difficult issue, so multiple-precision arithmetic was used in this work. As more harmonics are left in the solution, the relevant series converges to the exact solution exponentially fast.

Summarizing, we may say that the exact solution for the electric potential (and, consequently, for the current density distribution) in the elliptical Bitter disk was found by the Ritz method with multipoles as the appropriate function basis. The solution can be naturally extended over several disks and allows easy differentiation.

For identical disks, it is sufficient to calculate the coefficients of the series just once. Current density $(\ _1)$ in an elliptical Bitter disk creates magnetic field in point $_2$ equal (up to a constant factor) to the following integral over the volume of the disk:

$$=\int_V \frac{(\ _1)\times\ }{R^3}\ _1=\iint_V\left((\ _1)\times\nabla_{r_1}\left(\frac{1}{R}\right)\right)\ _1 , \tag{71}$$

where $=\ _2-\ _1$, $R=|\ \ |$. For a curl of a product of a scalar field and a vector field we have

$$\nabla_{r_1}\times\left((\ _1)\frac{1}{R}\right)=\frac{1}{R}\left(\nabla_{r_1}\times(\ _1)\right)-(\ _1)\times\nabla_{r_1}\left(\frac{1}{R}\right)=-(\ _1)\times\nabla_{r_1}\left(\frac{1}{R}\right), \tag{72}$$

as $\nabla_{r_1}\times(\ _1)=0$. The well-known formula for a volume integral of a curl yields

$$=-\int_V\nabla_{r_1}\times\left((\ _1)\frac{1}{R}\right)\ _1=-\int_S\frac{1}{R}\left(\ \ \times(\ _1)\right) . \tag{73}$$

The elliptical disk was defined above in Cartesian coordinates x_1, x_2, x_3. As component j_3 equals zero, component T_3 equals the integral over the lateral surfaces of the disk. It may be shown that

$$T_3 = \left(\int_{S_i} - \int_{S_o} \right) \frac{1}{R} |\ (\ _1)| \operatorname{sgn}(j_\varphi(\ _1)) dl dx_3 \quad , \tag{74}$$

where S_i is the inner lateral surface of the disk ($wx_1^2 + x_2^2 = p_1^2$, $-\frac{d}{2} \le x_3 \le \frac{d}{2}$), S_o is the outer lateral surface of the disk ($wx_1^2 + x_2^2 = p_2^2$, $-\frac{d}{2} \le x_3 \le \frac{d}{2}$), $\operatorname{sgn}(j_\varphi(\ _1))$ is the sign of the azimuthal component of the current density in polar coordinates, $dl = \sqrt{dx_1^2 + dx_2^2}$ is the element of length of the ellipse ($wx_1^2 + x_2^2 = p_1^2$ and $wx_1^2 + x_2^2 = p_2^2$, correspondingly). It should be taken into account that $j_3(\ _1) = 0$, the normal component of the current density vanishes on the surface of the disk, and the current density does not depend on x_3, so $(\ _1) = (x_1, x_2)$. Therefore, it is not difficult to perform integration with respect to x_3, obtaining

$$T_3 = \left(\int_{C_i} - \int_{C_o} \right) |\ (\ _1)| \operatorname{sgn}(j_\varphi(\ _1)) \frac{\ln\left| y_3 + \frac{d}{2} + \sqrt{\left(y_3 + \frac{d}{2}\right)^2 + (y_1 - x_1)^2 + (y_2 - x_2)^2} \right|}{\ln\left| y_3 - \frac{d}{2} + \sqrt{\left(y_3 - \frac{d}{2}\right)^2 + (y_1 - x_1)^2 + (y_2 - x_2)^2} \right|} dl, \tag{75}$$

where C_i and C_o are the ellipses $wx_1^2 + x_2^2 = p_1^2$ and $wx_1^2 + x_2^2 = p_2^2$, correspondingly, and y_1, y_2, y_3 are the Cartesian coordinates of $_2$ ($_2 = \{y_1, y_2, y_3\}$).

The integral over the bases of the disk yields the other two components of the field (T_1 and T_2):

$$\{T_1, T_2\} = \left(\int_{S_t} - \int_{S_b} \right) \frac{1}{R} \{j_2, -j_1\} dS \quad , \tag{76}$$

where S_t is the top base of the disk ($p_1^2 \le wx_1^2 + x_2^2 \le p_2^2$, $x_3 = \frac{d}{2}$), S_b is the bottom base of the disk ($p_1^2 \le wx_1^2 + x_2^2 \le p_2^2$, $x_3 = -\frac{d}{2}$). As the solution for current density is presented as a series in multipoles, the integral over the bases of the disk can be reduced to a sum of one-dimensional integrals (after integration with respect to r in the polar system of coordinates), but this was not done in this work, and the two-dimensional integral was calculated.

The suggested method has been verified in several ways. In particular, numerical tests showed that equations (11-12) and the boundary conditions for the potential and the current density in a single elliptical ring are satisfied with great accuracy. Two completely different methods were used for computation of magnetic fields from several elliptical disks, and the results agreed very closely. Comparison of the calculated magnetic fields for very narrow elliptical disks and elliptical wire coils yielded equally satisfactory results. Finally, calculated magnetic fields from disks with aspect ratio close to 1 (tilt angles $\ge 89^{\mathrm{O}}$) very closely agree with the fields obtained with the standard formulae for conventional Bitter coils [13].

It should be mentioned that so far we cannot explain as easily as we did for wire-wound coils the extremely high uniformity of the field that can be obtained with tilted Bitters (this uniformity was shown in our numerical experiments [4] discussed below) – the solution for current density in the disk is much more complicated than that for wire-wound tilted coils.

V Bitter Tilted Coil Magnet Design Example

The first design of a high field Bitter tilted coil magnet was discussed in work [4] by A.V. Gavrilin, M.D. Bird, et al. It was a three-coil magnet with the maximum purely transverse field of about 21T uniform within the whole aperture 38 mm in diameter. The three-coil configuration was suggested by M.D. Bird and optimized by A.V. Gavrilin [4]. The optimization made it possible to attain so high a field at only 13 MW power in the magnet. The magnet characteristics are given in Tables II and III [4].

Table II. Characteristics of three-tilted-coil Bitter magnet.

Total number of coils	3
Material	Copper
Inner radius (mm)	19
Outer radius (mm)	305
Overall length (mm)	1195
Rated current (kA)	40
Maximum field @40 kA	20.8
Total power in the magnet (MW)	12.62
Total number of turns	274

Table III. Characteristics of coils of three-tilted-coil Bitter magnet.

Coil #	1	2	3
Inner radius (mm)	19	77	152
Outer radius (mm)	74	149	305
Length (mm)	440	721	1195
Number of turns	94	80	100
Turn thickness (mm)	2.2	3.75	4.15
Tilt angle (deg.)	-45	+45	+45
Current (kA)	+40	-40	-40
Peak current density (A/mm^2)	647	188	82
Average current density (A/mm^2)	331	148	63
Peak stress (MPa)	~400	---	---

High-accuracy calculations showed that the transverse field that can be obtained by means of the magnet is amazingly transverse and uniform over the whole bore along the length of at least 150 mm (Fig. 10). It is impossible to attain such a large zone of uniformity at such modest power in the magnet using other known approaches. - Actually, a standard configuration for a transverse field Bitter magnet is split pairs [4], where two solenoids are assembled end-to-end with a gap between them. Such a design makes it possible to obtain transverse fields higher than 20T, in a relatively wide gap, at reasonable power in magnets [2], [4]. However, these fields are essentially non-uniform.

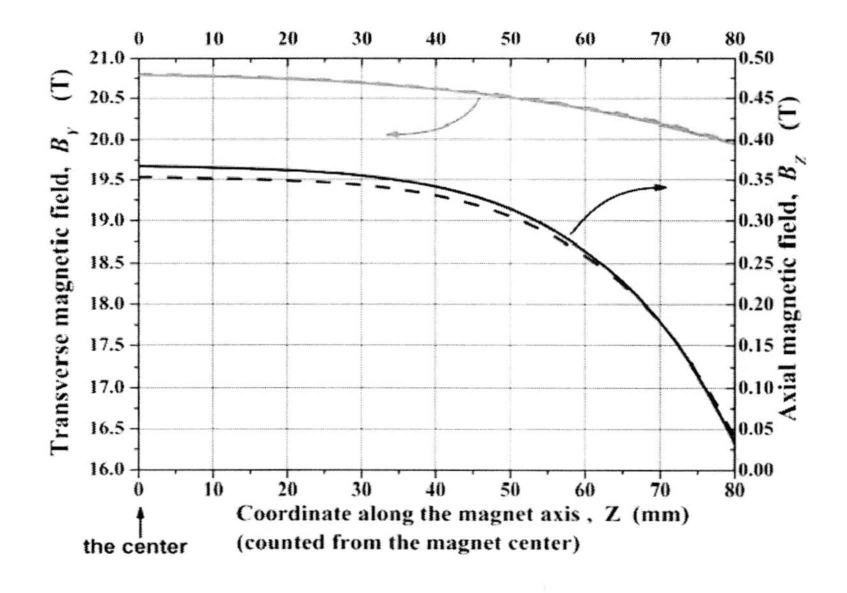

Fig. 10. Magnetic field distribution along the three-tilted-Bitter-coil magnet. B_Y is the desirable transverse component, B_Z is the axial (solenoidal) field. Solid curves – the fields on the magnet central axis Z (X=0,Y=0); dashed – on a line parallel to Z and going through the point X=9.9mm,Y=9.9mm, i.e., at a radius of 14 mm (5 mm away from the winding inner surface).

A finite element elastic stress analysis through ANSYS was also conducted [4], which revealed rather moderate, i.e., acceptable, stresses in the coils [4] (Table III, Fig. 11) with the maximum in the innermost coil in the horizontal, *XZ*, plane. What is more interesting, the analysis showed complex patterns of displacement of tilted coils under Lorentz forces [4], and these patterns are rather typical and showy. The point is that, although a Bitter tilted coil is still a solenoid, the force distribution and direction in it are very specific and completely different from those in a conventional, 'non-tilted', Bitter coil. While in the latter the Lorentz forces are radial, creating mostly tangential stress, in a tilted Bitter coil magnet the Lorentz forces are rather axial than radial, at least in the innermost coil where the field is highest. These forces mainly try to rotate the coil in the vertical, *YZ*, plane around the coil center (Fig. 12), albeit not only this, creating rather specific compressive-bending-shearing stress-strain patterns in the disks.

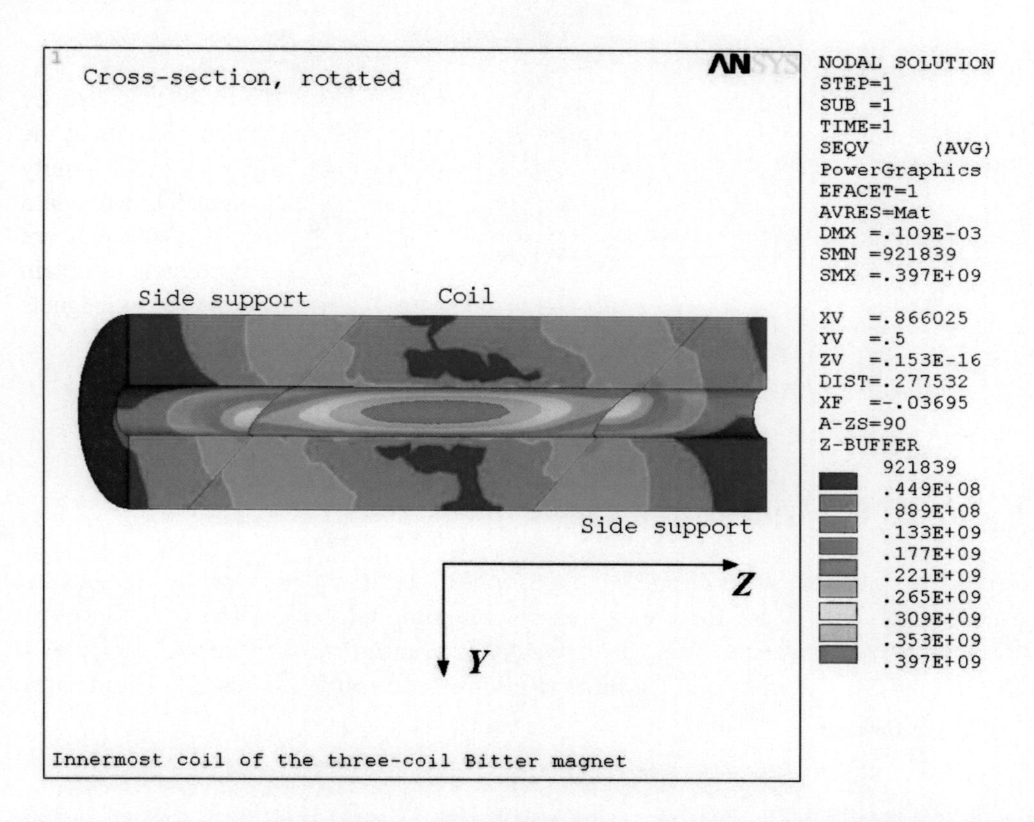

Fig. 11. Von Mises stress distribution within the innermost coil of the three-tilted-Bitter-coil magnet at the rated current (20.9T maximum field, 12.7MW power in the magnet) [4]. Cross-sectional view. 3D calculation results of elastic stress analysis through ANSYS. The stress peak value of ~400 MPa occurs on the inner surface of the coil around the middle, in *XZ*-plane.

Lorentz forces of a similar character should be expected in wire-wound tilted coils, too. This imposes heavy demands on the support structures, especially in superconducting high-field magnets. In fact, if a tilted helix form (Fig. 3) is used, the rotating Lorentz forces are in essence applied directly to the form.

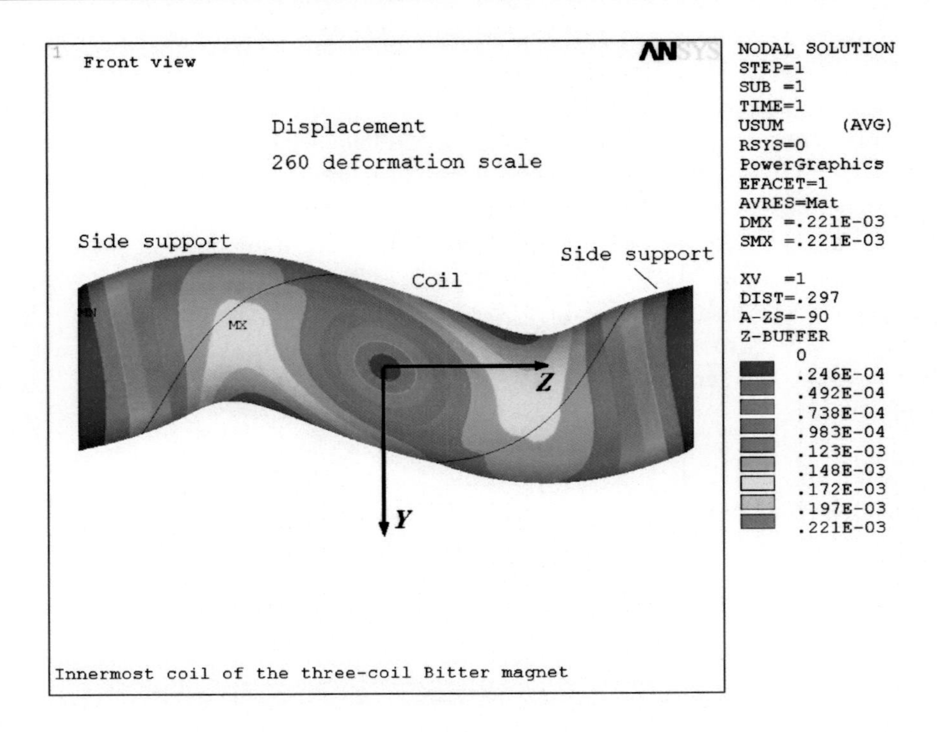

Fig. 12. Displacement distribution on the surface of the innermost coil of the three-tilted-Bitter-coil magnet at the rated current. Typical 3D deformation of the coil (260 deformation scale) under Lorentz forces, computed through ANSYS. Side view. The edges (end-planes) of the side support structures are fixed.

VI Confocal Elliptical Bitter Disks

In the previous sections, we considered tilted coil Bitter disks bounded by homothetic ellipses. However, this is not the only geometry admitting an exact solution. A different geometry (Fig. 13) is also analyzable, where the area is bounded by two confocal, rather than homothetic, ellipses. The inner and the outer ellipses are described in Cartesian coordinates x, y by equations:

$$\frac{x^2}{a_i^2}+\frac{y^2}{b_i^2}=1, \tag{77}$$

where $i=1$ for the inner ellipse, $i=2$ for the outer ellipse, and $a_1^2-b_1^2=a_2^2-b_2^2=h^2$ (the ellipses have common foci in points ($\pm h$, 0)). Again, we seek the partial solution $\Phi(x,y)$ of the equation $\Delta\Phi=0$ in the area limited by the two confocal ellipses with a cut along the abscissa axis (in the left half-plane). On the upper and the lower banks of the cut, the function equals π and $-\pi$, correspondingly. At the elliptic boundaries, the projection of the gradient of this function on the normal to the ellipses equals zero. It is common knowledge that the variables in the Laplace equation can be separated in the elliptical coordinates. Therefore, let us introduce the special elliptic coordinates u, v using the following formulae [16]:

$$x = h\cosh(u)\cos(v),$$
$$y = h\sinh(u)\sin(v). \quad (78)$$

If we introduce complex variables $z = x + iy$ and $w = u + iv$, then

$$z = h\cosh(w). \quad (79)$$

This formula defines a one-to-one conformal mapping of the rectangle $0 < u_1 \le u \le u_2$, $-\pi < v \le \pi$ in the complex plane w into the area between the confocal ellipses with a cut in the complex plane z, if $h\cosh(u_i) = a_i$, $i = 1,2$. The inverse conformal mapping is defined by the following formula:

$$w = \ln\left(\frac{z}{h} + \sqrt{\left(\frac{z}{h}\right)^2 - 1}\right). \quad (80)$$

To remove the ambiguity, we choose such value of the square root that $\left|\frac{z}{h} + \sqrt{\left(\frac{z}{h}\right)^2 - 1}\right| \ge 1$ and such value of the logarithm that $-\pi < \mathrm{Im}(w) \le \pi$. Then the following function yields the sought-for solution:

$\Phi(x, y) = \Phi(z) = \mathrm{Im}(w(z)) = v$. Indeed, this function satisfies the Laplace equation as it equals the imaginary part of an analytical function. The function equals $-\pi$ and π on the lower and the upper banks of the cut, respectively. It is also almost evident that the projection of the gradient of this function on the normal to the ellipses equals zero. However, we shall calculate the gradient explicitly, as it equals the current to a constant factor. In the curvilinear orthogonal coordinates u, v we have [16]:

$$\nabla v = \frac{1}{\sqrt{g_{uu}}}\frac{\partial v}{\partial u}\mathrm{i}_u + \frac{1}{\sqrt{g_{vv}}}\frac{\partial v}{\partial v}\mathrm{i}_v, \quad (81)$$

where

$$g_{uu} = g_{vv} = h^2(\sinh^2(u) + \sin^2(v)), \quad (82)$$

$$\mathrm{i}_u = \frac{1}{\sqrt{g_{uu}}}\left(\frac{\partial x}{\partial u}\mathrm{i} + \frac{\partial y}{\partial u}\mathrm{j}\right),$$

$$\mathrm{i}_v = \frac{1}{\sqrt{g_{vv}}}\left(\frac{\partial x}{\partial v}\mathrm{i} + \frac{\partial y}{\partial v}\mathrm{j}\right), \qquad (83)$$

and the orthonormal vectors i, j are directed along the axes x and y, respectively. Finally,

$$\nabla v = \frac{1}{\sqrt{g_{vv}}}\mathrm{i}_v = \frac{-\cosh(u)\sin(v)\mathrm{i} + \sinh(u)\cos(v)\mathrm{j}}{h(\sinh^2(u) + \sin^2(v))}. \qquad (84)$$

It is easy to check that the gradient is tangential to the confocal ellipses.

The solution can be trivially adapted to a case where the cut is made along the lesser semi-axis.

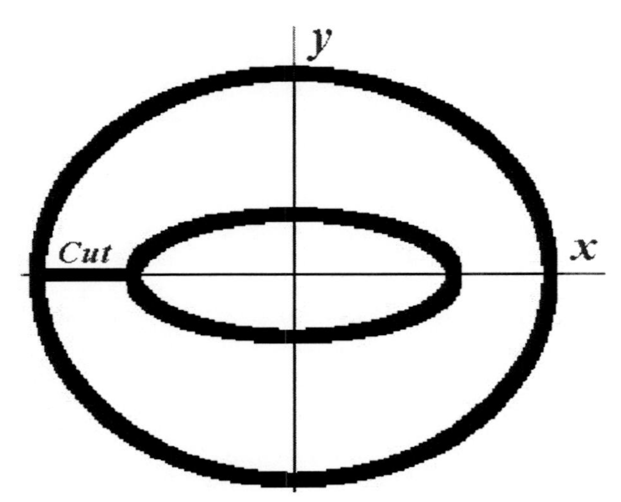

Fig. 13. Sketch of a disk bounded with confocal ellipses with a coordinate system aligned with it.

Although the solution is much simpler than that for homothetic ellipses, the downside is, evidently, that either the outer or the inner radius of the relevant tilted coil is not round. That means that either the bore or the bandage of the magnet is not round. This drawback can be partly mitigated if the coils comprising the magnet are tilted at different angles. In addition, the field uniformity that can be obtained with confocal elliptical Bitter disks is yet to be studied.

VIIWinding Technology

While, as mentioned above, the manufacturing of tilted Bitter coils seems to be rather similar to that of conventional Bitter solenoids, this is not completely the case for wire-wound tilted coils. Several technologies can be used to manufacture tilted wire-wound coils so as to solve the major problem of holding tilted turns in a predetermined 'tilted' position [4], [6], [7]. One possibility, which looks the most natural and straightforward if not obvious, is to use a tilted helix form that can be easily fabricated by first milling a groove of the correct geometry into a tube (Fig. 3). A commercially available numerically controlled machining software package, such as FeatureCAM, may be used to generate the tool cutting paths directly from the path equation (see (1)), with specified parameters of diameter, pitch, tilt angle and number of turns (Fig. 14). The conductor is then wound into the groove (Fig. 15), and any necessary overwrap or fill materials are applied. The wound coil forms are then nested and aligned to one another to form the complete magnet assembly. Spatial position accuracy of +/- 0.1 mm may be achieved by this method. The technology used by AML, Inc [7], [17], which was the first US company to start manufacturing tilted coils, looks similar to the described one. Seemingly, it can provide sufficient strength and integrity at the field of a few teslas; in so doing the ultimate in the average current density can be reached if the gap between adjacent turns (see (5)) is held minimal. At higher magnetic fields (~ 5-8 T and more), a tilted helix form, typically made of non-metallic (i.e., relatively weak) material such as G10 composite, along with epoxy resin matrix (if the coil is epoxy-impregnated), might be damaged by the conductor exhibiting torque due to the rotating Lorentz forces (see Section V. Bitter Tilted Coil Magnet Design Example). If so, a non-metallic tilted helix form should be substituted with one made of a stiffer material, such as stainless steel or another metallic alloy, though, perhaps, some modern ceramics can turn out to be appropriate, too.

Fig. 14. Screen shot of tilted helical groove milling operation programmed in FeatureCAM.

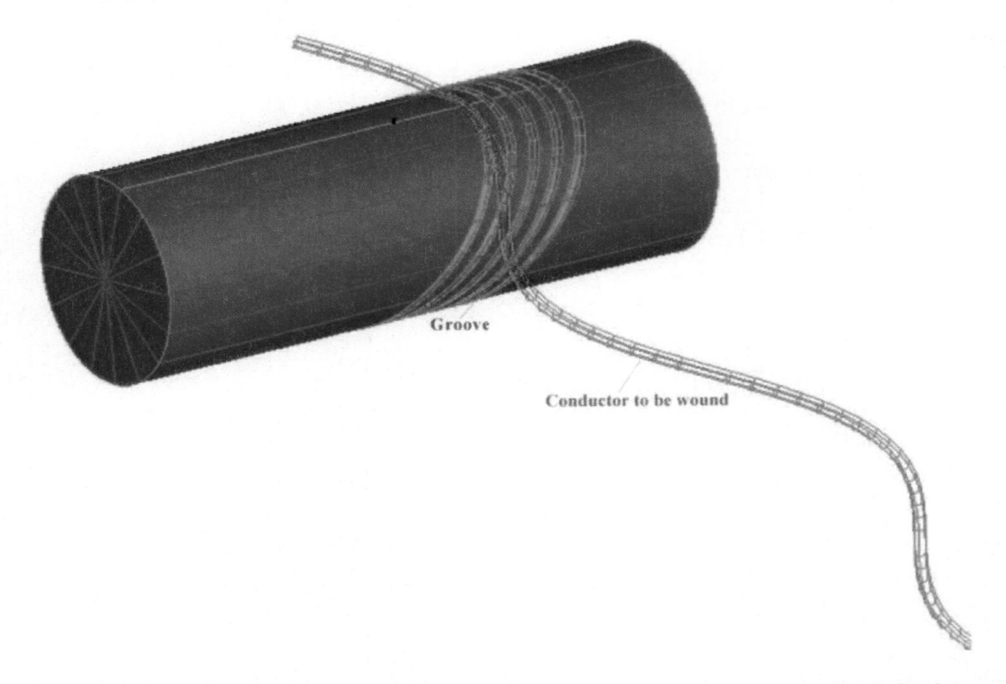

Fig. 15. Screen shot of tilted helix path with a conductor to be wound – put into the helical groove (programmed in FeatureCAM).

Another solution of the problem of holding tilted turns in a predetermined position consists of flat ellipse-shaped pancake type windings instead of solenoidal ones. A long tilted coil can be assembled from a proper number of elliptical flat pancakes (or double pancakes) tilted and stacked at any reasonable angle off the central axis instead of layerwise winding the coil in a long solenoidal shape [6]. Manifestly, the idea is 'accordant' to the one of tilted Bitter coils. Each pancake has its own side support ('inter-pancake support', Fig. 16), and side and inner supports for the whole coil are used as well. All the pancakes are electrically connected in series. Every elliptical flat pancake representing a flat spiral can be wound using a rather traditional pancake winding technology. Obviously, this technology can only be used for the manufacturing of radially thick tilted coils. For a tilted multi-pancake coil, a rectangular conductor looks preferable to a round one. A superconducting cable-in-conduit conductor may be a good candidate for a tilted superconducting coil, as the conduit can provide the whole ellipse-shaped pancake with needed strength and stiffness.

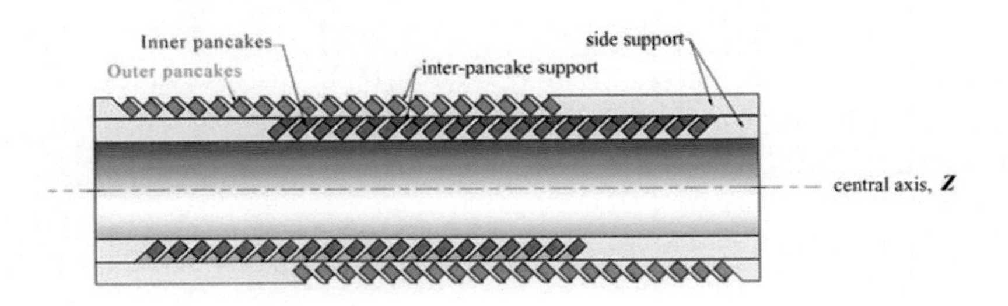

Fig. 16. Sketch of the simplest tilted coil magnet assembled from flat pancake-type elliptical coils. Cross-sectional view.

VIII Conclusion

The practical potential of tilted coil magnets is not fully understood yet. However, we do believe that tilted coil magnets can be used for a diversity of experiments in many areas of science. Actually, research and development work is still in its beginning, though considerable progress has already been made. It takes time to adapt the new concept to needs of potential users.

Let us summarize the contents of this chapter:

1). The mathematical foundation is laid for a relatively new type of magnets for generation of a uniform transverse magnetic field in a large volume of space - tilted coil magnets.

2). It is proven that the wire-wound tilted coils have the ideal distribution of the axial current density, "cosine-theta" plus constant.

3). Magnetic fields calculated for a tilted Bitter coil magnet were shown to be highly uniform.

4). A finite element stress analysis showed complex patterns of deformation of tilted coils under Lorentz forces.

5). Manufacturing technologies for tilted coils are discussed.

References

[1] Wilson, M.N. *Superconducting Magnets*; Clarendon Press: Oxford, 1983; ch. 3.

[2] Bird, M.D., Dixon, I.R., Toth, J. *IEEE Trans. Appl. Supercond.*, 2004, vol. 14, NO. 2, 1253-1256.

[3] Meyer, D.I., Flasck, R. *Nuclear Instruments and Methods,* 1970, NO. 80, 339-341.

[4] Gavrilin, A.V., Bird, M.D., Bole, S.T., Eyssa, Y.M. *IEEE Trans. Appl. Supercond.*, 2002, vol. 12, NO. 1, 465-469.

[5] Devred, A. *Proc. of 2003 Particle Accelerator Conference, 0-7803-7739-9* / 2003 *IEEE*, **146**-150.

[6] Gavrilin, A.V., Bird, M.D., Keilin, V.E., Dudarev, A.V. *IEEE Trans. Appl. Supercond.*, 2003, vol. 13, NO. 2, 1213-1216.

[7] Goodzeit, C.L., Ball, M.J., Meinke, R.B. *IEEE Trans. Appl. Supercond.*, 2003, vol. 13, NO. 2, 1365-1368.

[8] Goodzeit, C.L., Meinke, R.B., Ball, M.J. *IEEE Trans. Appl. Supercond.*, 2003, vol. 13, NO. 2, 2235-2238.

[9] Nguyen, D.N., Sastry, P.V.P.S.S., Zhang, G.M., Knoll, D.C., Schwartz, J. *IEEE Trans. Appl. Supercond.,* 2005, vol. 15, NO. 2, 2831-2834.

[10] Gavrilin, A.V., Bird, M.D. *US Patent* NO. **6**,876,288, April 2005.

[11] Parrel, J.A., Zhang, Y., Field, M.B., Cisek, P., Hong, S. *IEEE Trans. Applied Supercond.*, 2003, vol. 13, NO. 2, June , pp. 3470-3473.

[12] Iwasa, Yu. *Case Studies in Superconducting Magnets. Design and Operational Issues;* Plenum Press: New York, 1994, ch. 8.

[13] Montgomery, D.B. *Solenoid Magnet Design*; Wiley & Sons Press, 1969.

[14] Rektorys, K. *Variational Methods in Mathematics, Science and Engineering*; Reidel: Dordrecht/Holland e.a., SNTL: Prague, Czechoslovakia, 1980.
[15] Miller, W., Jr. *Symmetry and Separation of Variables;* Addison-Wesley: Reading, Massachusetts, 1977, section 1.2., case II.
[16] Korn, G.A., Korn, T.M. *Mathematical Handbook for Scientists and Engineers*; McGraw-Hill, 1968.
[17] Goodzeit, C.L., Meinke, R.B., Ball, M.J. *US Patent* NO. **6**,921,042, July 2005.

In: Superconductivity, Magnetism and Magnets
Editor: Lannie K. Tran, pp. 173-190
ISBN: 1-59454-845-5

Chapter 6

AC MAGNETIC SUSCEPTIBILITY RESPONSE

Ozer Ozogul
Ankara University, Faculty of Science, Department of Physics,
Tandogan, 06100 Ankara, Turkey

Abstract

AC magnetic susceptibility measurement is perhaps the most useful experimental technique for understanding the magnetic properties of superconductors. This measurement technique includes an in phase (or real) component χ_n' and an imaginary out of phase component χ_n'' (n=1, 2, 3...). The AC magnetic susceptibility considered here is in principle deduced by means of a phase sensitive detector. The phase sensitive detector is operated using a filter to accept the fundamental and higher harmonic susceptibilities and no filtering is used for AC magnetization hysteresis loops. The concept of a critical-state model for superconductor and its use in the interpretation of AC magnetic measurement in terms of critical-current density were introduced by Bean in which critical current density was assumed to be independent of the local magnetic field. Further critical-state models have been proposed, with a different local magnetic field dependence of critical current density. Nevertheless the observed external parameters (AC field amplitude and frequency, dc magnetic field, temperature etc.) of fundamental and higher harmonics can not be described within any critical-state model. In order to explain such dependencies, creep models have been introduced, such as Giant Flux Creep, Vortex Glass and Collective Creep.

In this chapter, the AC magnetic susceptibility technique and critical-state models will be briefly reviewed. The experimental AC susceptibility data on some high T_c superconductors will be compared with several theoretical models.

1 Magnetic Measurement Techniques

The critical current density of the superconductors has been usually obtained from direct *V-I* measurements. This technique requires the attachment of electrical leads to the sample material under study and this is not always convenient. Alternatively, the magnetic measurement techniques are very commonly used as the measurements require no electrical contacts and can be performed on a sample of small size. In order to employ magnetic measurements in the determinations of critical current density, a suitable critical-state model must be invoked. The existing family of magnetic measurements techniques, based on recording the voltage generated in detection coil, can probe an entire sample volume and therefore it provides us with a volume average of the sample's magnetic response.

1.1 DC Technique

The starting point is the magnetic moment, m, to describe the measurement techniques. Magnetic moment resulted from permanently magnetized sample or by magnetizing samples through the applications of an applied magnetic field, H, is a measure of the magnetic field generated by the sample itself. The volume magnetization is defined as $M = m/V$, where V is the sample volume. The magnetic susceptibility relating the magnetization to the applied field is given by $\chi = M/H$. In a DC measurement system, such as vibrating sample magnetometer (VSM) or SQUID magnetometer, a magnetic moment is measured by using an applied DC field (H_{dc}). In order to detect the change in magnetic flux due the presence of the magnetic moment, a detection coil is used. Since the applied magnetic field is constant, the magnetic flux change within detection coil is generated by moving the sample. In order to move the sample, either the sample is vibrated near the detection coil or the sample is passed through the detection coil. The output signal as a function of time is detected but the magnetic moment of the sample does not change with time and therefore it does not represent the AC response of the sample itself. The DC magnetic susceptibility is a measure of sample's magnetization and it is defined as, $\chi_{dc} = M/H_{dc}$.

1.2 AC Technique

The AC magnetic technique uses a detection coil as that in the DC ones, but the sample movement is not required. However the sample movement is used to increase measurement accuracy. In this technique the sample is centered within a detection coil and driven with an applied AC field, $H = H_{ac} \cos(wt)$, where w is the $2\pi f$. The magnetization of the sample periodically changes in response to an applied AC field. The output signal is proportional to the change in magnetic moment (Δm). Thus, the AC susceptibility gives an indication of the slope ($\chi_{ac} = dm/dH$) of the magnetization curves. Thus AC response of the magnetic system can be studied. When a sample is exposed to an external AC field, the response of the sample may not necessarily be a pure sine response. This is due to their pinning properties, which govern the dynamic response to a sinusoidal field. In general AC susceptibility describes the magnetization $M(t)$ of a sample as a function of the AC magnetic field. The time dependent magnetization in that case can be expressed as a sum of harmonics, which oscillate at harmonics of ac field frequency,

$$M(t) = H_{ac} \sum_{n=1}^{n=\infty} [\, \chi_n' \cos(nwt) + \chi_n'' \sin(nwt) \,] \tag{1}$$

where χ_n' and χ_n'' represent the components of the AC magnetic susceptibility which are in-phase (real) and out-of-phase (imaginary) respectively and n is the order of harmonic components. The real part χ_1' corresponds to the dispersive magnetic response and the imaginary part χ_1'' corresponds to energy dissipation. Figure 1 illustrates the AC field response of superconducting materials with nonlinear magnetic behavior. Figure 1(a) shows the odd harmonics components (n=1,3,5,7) and sum of these harmonic components as a function of time. The magnetization which is sum of harmonic components (n=1,3,5,7) as a function of magnetic field is given by Fig.1b.

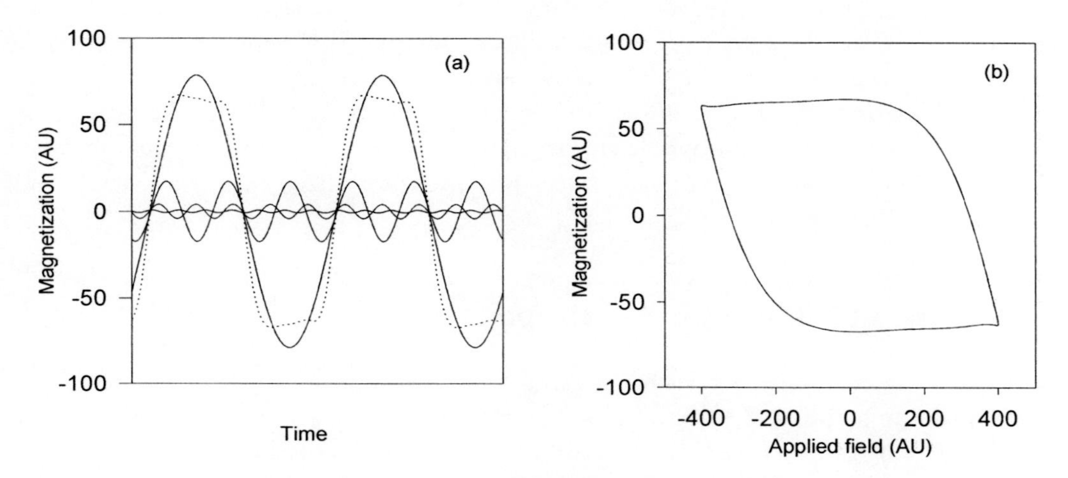

Figure 1 a) Schematic illustration of odd harmonic components (solid line) and sum of these components (dot line), b) The magnetization that is sum of harmonic components as a function of magnetic field.

In complex notation, χ_n' and χ_n'' are combined to form the complex susceptibility, $\chi_n = \chi_n' + i\chi_n''$ (n=1,2,3...). The AC magnetic susceptibility considered here is in principle deduced by means of a phase sensitive detector (PSD). In many experimental systems, the phase angle θ between the sample signal and PSD may be unavoidable. This means that, the phase angle of the PSD used to detect the coil voltage must be adjusted to correctly separate the real and imaginary parts of susceptibility. The proper separation of χ_n' and χ_n'' requires that the phasing be performed with either a test sample with a known $\chi_1''=0$ or under measurement conditions where $\chi_1''=0$ for the sample under study. The superconductor at low temperature and low measuring field can be used as a test sample. The phase-sensitive signal V_θ measured at θ will be proportional to real component and the signal $V_{\theta+90}$ measured at $\theta+90^o$ will be proportional to imaginary components. The real and imaginary components of the measured susceptibility are determined from the following relationship,

$$\chi_n' = \frac{\alpha V_\theta}{VnfH_{ac}},$$

$$\chi_n'' = \frac{\alpha V_{\theta+90}}{VnfH_{ac}}, \qquad (2)$$

where α is calibration coefficient, V is the sample volume and f is frequency of AC field. In the measurements of the fundamental susceptibility, the PSD uses a low-pass filter to reject higher harmonics and output signal to yield a better measurement of χ_1' and χ_1'', since the output signal appears at the fundamental excitation frequency or bears a harmonic relationship to it. In the harmonic measurements, the PSD was fed with an external reference frequency of nf instead of the basic excitation. For this reason the phase sensitive detector is operated using a band-pass filter to accept the higher harmonic susceptibilities and no filtering is used for AC magnetization hysteresis loops. In this chapter, AC measurements were carried out by using the Lake Shore AC susceptometer 7130 series with closed cycle refrigerator. In the measurements, the phase angle of the PSD was adjusted for each frequency, such that $\chi_1''= 0$ at the lowest temperature for the lowest field, and the same phase adjustment was maintained for measurements carried out at higher field. In order to make quantitative comparison with the model calculations, we normalized χ_n to χ_1' at the lowest temperature and lowest field.

2 Outline of Models for AC Response

AC magnetic susceptibility is perhaps the most useful experimental technique to understand the magnetic properties and to estimate the critical current density by use of critical-state models. Nevertheless, it is an indirect measurement, so it is necessary to introduce various models to interpret experimental results. The critical-state model assumes that, for $H > H_{c1}$, the penetrated supercurrent flows with a density J whose magnitude equals to the critical-current density $J_c(B)$, where B is the local internal field. The entire or a part of the superconductor where $|J|= J_c$ is said to be in the critical-state [1]. The critical state describes a current distribution throughout the superconductor that is supposed to be determined by the effectiveness of material inhomogeneities in pinning a flux distribution. Several models have been proposed to interpret the magnetic behavior, with a different local magnetic field dependence of critical current. Some of these models are given Table1. In both the theoretical and experimental investigations, Bean and Anderson-Kim models have been extensively used.

Table 1 Several critical-current models

Model	Critical-Current	References
Bean	$J_c = H_p / a$	[1, 2]
Anderson-Kim	$J_c = k / (B_o + B)$	[3-6]
Exponential-Law	$J_c = A \exp(-B / C)$	[7]
Linear	$J_c = A - C B$	[8]
Power-Law	$J_c = k B^{-q}$	[9]

2.1 Bean Model

The first model taking into account higher harmonics of the AC magnetic susceptibility was introduced by Bean [1,2] in which critical current density was assumed to be independent of the local magnetic field. According to Bean, the field profile interior to the sample is a linearly decreasing function of distance from the edge to the inside of the sample. As the field is increased, the magnetization will decrease as the normal regions containing the vortex cores grow at the expense of superconducting regions. When the field is decreased, the current flows in the opposite direction and the magnetization becomes positive. This model attributes the harmonic generation to the hysteretic. Nevertheless only odd harmonics are predicted, because of the symmetric hysteresis loops. Bean's model gives analytic expressions for the magnetization, M, as a function of the applied magnetic field [2, 10]. The equations for susceptibilities as a function of field are derived from Fourier transform of the magnetization [11,12]. The model calculations for M-H loops and susceptibilities (n=1 and 3) are plotted in Fig.2. The magnetization loops shown in Fig.2(a) were calculated with various ac field amplitudes. Figure 2(b) shows that the fundamental and third harmonic susceptibilities as a function of AC field amplitude are calculated. In order to obtain model calculation as a function temperature, we need to know the temperature-dependent of J_c(T). There is another point worth mentioning that the height of the $|\chi_n|$ peak is independent of the amplitude H_{ac} and its frequency within the Bean model.

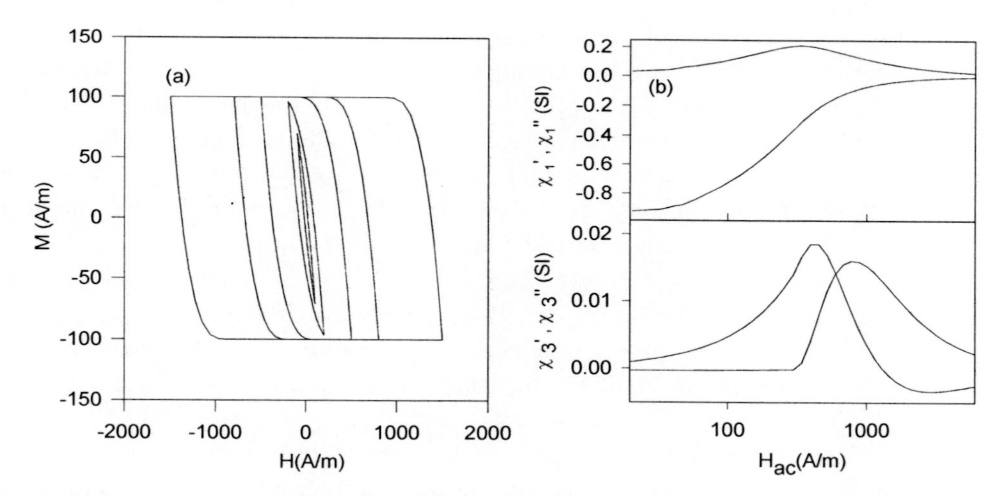

Figure 2 a) M-H loops with various AC field amplitudes, b) First and third harmonic susceptibilities as a function of AC field amplitude.

On the other hand the critical current density J_c can be estimated from both AC and DC magnetic measurements. Critical current density is proportional to the width ΔM of the hysteresis loops in the DC measurements. In the AC measurements, the fact that penetration field is equal to the amplitude of AC field at the peak in the imaginary fundamental susceptibility. Therefore, critical current density is proportional to the H_{ac} /a where a is sample dimension.

2.2 Anderson-Kim Model

Unlike the original Bean critical-state model, which predicts only odd harmonics even in the presence of nonzero DC magnetic fields, the local field dependence of critical current density proposed by Anderson and Kim *et al.* [3-6] predicts even harmonics in a slight DC field. This case will be called "Anderson-Kim model". The Anderson-Kim critical-state model assumes a critical current density J_c that decreases with increasing local field B_i , $J_c = k / (B_o + |B_i|)$ where k and B_o temperature dependence parameters [13,14]. Ji *et al.* [10] point out that any field dependence in J_c will lead to generation of both even and odd harmonics. In the critical-state model, the Lorentz force per unit volume is balanced by the pinning force density. In comparison with the critical-state models, the Anderson-Kim model is only a simple example with which the essence of the underlying physics in the high temperature superconductor (HTS) can be seen. Let us consider an infinite slab or infinitely long cylinder sample and the external magnetic field parallel to the slab or the cylinder axis is applied. Using Ampere's law and Anderson-Kim critical-current density, we have

$$\left|\frac{dB_i}{dx}\right| = \frac{k}{B_o + B_i}. \tag{3}$$

where, x is the coordinate perpendicular to the slab or the radial coordinate of a cylinder. Equation 3 for the boundary condition $B_i = H = H_{dc} + H_{ac}\cos(wt)$ can be solved. If H increases, x decreases from surface ($x = a$) inward and reach zero. When x equals zero, the sample is completely penetrated. This field is called the full penetration H_p. In this range, variations of x can be determined as $x = a - [(B_o+H)^2 - B_o^2] / 2k$. When $x=0$, H is equal to the H_p which is obtained as $B_o[(1+p^2)^{1/2} - 1]$ where $p = (2ka)^{1/2} / B_o$. Chen and Goldfarb [11] have made derivations of magnetization dependent on p for an orthorhombic specimen with a rectangular cross section $2a$x$2b$ ($a \le b$). According to them, the shape of magnetization curves is determined by p. However they have not taken into account H_{dc} field. Ji *et al* [10] used the simplified model, in which B_o was taken to be zero (for limit $p \to \infty$), and derived equations for the magnetic hysteresis loop around a DC bias field for a slab geometry. But, the magnetization used in their study has not included variation of p. Müller [15] calculated the temperature and field dependencies of the fundamental susceptibility using Anderson-Kim model. Ozogul and Aydinuraz [16] concluded that, the parameter p is related to the J_c nonuniformity within the sample and characterizes the shape of the harmonic susceptibilities.

The flux density inside the superconductor follows directly from Eq. (3) and then for an infinite slab or infinitely long cylinder, the magnetization is given by

$$M(t) = \frac{2}{a^2}\int_0^a xB(x)dx - H(t). \tag{4}$$

Where a is the dimension of slab or the radius of cylinder. Figure 3 shows the *M-H* loops derived from Anderson-Kim model for $p = 0.1$ and 1000. For $p = 0.1$, the curves are very similar to those obtained from Bean's model. On the other hand, for $p=1000$ is almost the same as for limit $p \to \infty$ where Anderson-Kim model reduced to simplified Kim's model [10].

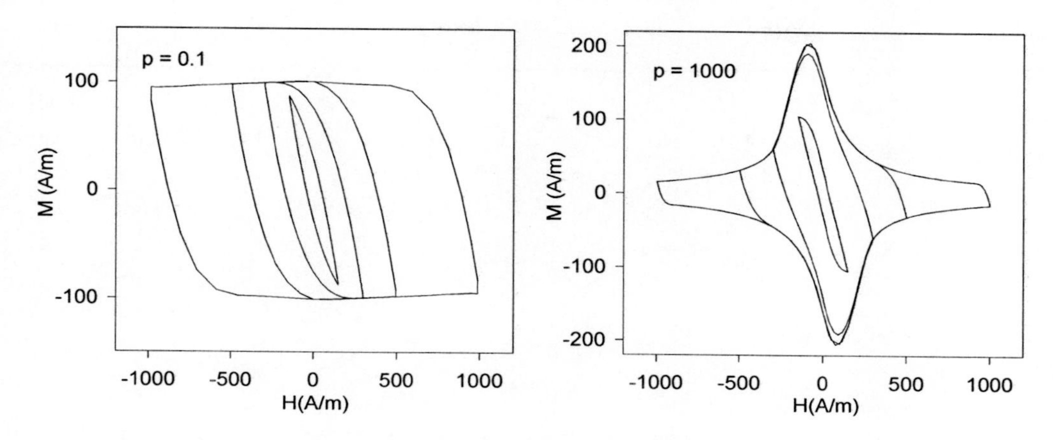

Figure 3 AC Magnetization loops obtained from Anderson-Kim model for p=0.1 and 1000. In the model calculations, penetration field is taken as 300 A/m.

The complex ac susceptibility, $\chi_n = \chi_n' + i\chi_n''$, can be derived by means of the Fourier transform. The explicit forms of χ_n' and χ_n'' are given by,

$$\chi_n' = \frac{1}{\pi H_m} \int M(t) \cos(nwt)\, d(wt)$$

$$\chi_n'' = \frac{1}{\pi H_m} \int M(t) \sin(nwt)\, d(wt). \tag{5}$$

The fundamental susceptibilities as a function of applied field were calculated different p values. In order to see more clearly p parameter dependence, it is useful to plot χ_n'' as a function of χ_n' (Cole-Cole plot). Figure 4(a) displays the Cole-Cole plot of first harmonic. The Cole-Cole plots of the first harmonics are characterized by a dome-shaped. As can be seen in Fig 4(a), the height of the peak in χ_1'' depends on the p parameter. In other words, when p is increased, the imaginary component of fundamental susceptibility increases. Prediction of the model in the value $\chi_{1,max}''$ is 0.212 (for p= 0.1) like that in the Bean model and 0.401 (for p= 1000) very close to the Diffuse [17] model prediction (0.417). In the Diffuse model, the apparent susceptibility of a superconductor arising from shielding is due to eddy currents. On the other hand, the effect of the parameter p on the third harmonic susceptibility can be clearly seen from Fig. 4(a), where plots χ_3'' as a function of χ_3' are plotted. The polar plots turn in the direction of clockwise and tend to occupy the right semi-plane when p is decreased. In addition of these evident changes, the area of the Cole-Cole plots is not constant as p value varies.

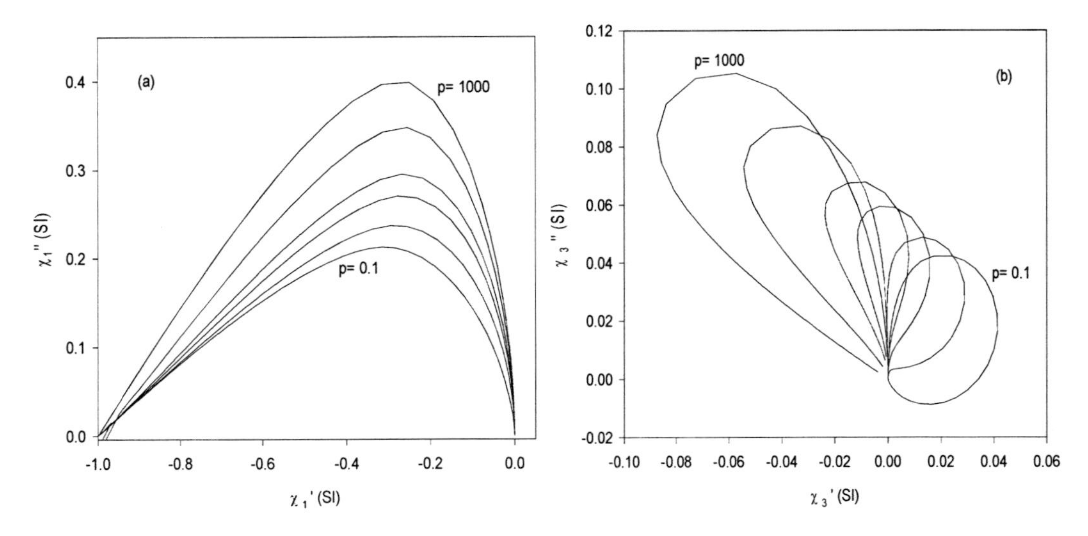

Figure 4 Cole-Cole plots for different p values a) Fundamental susceptibilities b) Third harmonic susceptibilities.

For the lowest p value, the polar plots reveal the same behavior as expected Bean's model. However the behavior of the Cole-Cole plots of third harmonic deviates from the Bean model when p value is increased. According to Ozogul and Aydinuraz [16], the harmonic susceptibilities for lower p values correspond to the similar characteristic of magnetic behavior of superconductors which possess intergranular interaction while the harmonic susceptibilities for higher p values correspond to the behavior of homogeneous superconductors such as single crystal or samples with contiguous oriented grains. They suggested that the physical and magnetic properties of the sample determine the p parameter.

Anderson Kim model has been considered to investigate the temperature dependence of AC magnetic characterization of the high temperature superconductor in the presence of either a pure ac field or with a superimposed DC field [15, 18-21]. In order to compare model calculation with temperature-dependent experimental measurements, we need to know the temperature-dependent parameter H_p. The general form for H_p has been defined as [22],

$$H_p(T) = H_p(0)\left(1-\frac{T_p}{T_c}\right)^{\beta} \tag{6}$$

where T_p is the peak temperature corresponding to $\chi_{1,max}''$. It assumed that, the fundamental out of phase susceptibility has its peak value at T_p when the ac field amplitude, H_{ac}, becomes equal to the penetration field, H_p. We make fit T_p- H_{ac} experimental data in order to obtain penetration field as a function of temperature. On the other hand, Anderson [3] proposed that B_o corresponds to the minimum magnetic flux density of a flux bundle, although the bundle size possibly fluctuates and this makes difficult to confirm his proposal. In fact, B_o is assumed as temperature-dependent parameter in the Anderson-Kim model and physically correlated to irreversibility field [13, 14].

3 Granular Superconductors

Many measurements have indicated that high temperature superconductors produced by sintering consist of superconducting grains. The magnetic properties and the current density in granular superconductors are governed not only by the nature of the diamagnetic grains but also by their interconnections which constitute the superconducting matrix [18, 23, 24]. In these materials, the grain boundaries act as Josephson junctions that are interconnected and form a complicated weak-link. Applications of a small magnetic field ($H < H_{c1,m}$) to such a granular superconductor will induce screening currents that flow around the outer surface of the sample. For applied field greater than the Josephson critical field ($H_{c1,m} < H < H_{c2,m}$), intergranular vortices (but no intragranular vortices) will nucleate along grain boundaries at the surface and then penetrate the weak-link network. This gives rise to macroscopic current loops (extending over the whole sample) as well as to microscopic loops of London currents circulating around individual grains. At fields larger than the weak link decoupling field ($H_{c2,m} < H < H_{c1,g}$), the macroscopic current vanishes and the magnetic behavior is governed solely by the London currents surrounding decoupled grains [25]. When the applied field is greater than the lower critical field for grains ($H > H_{c1,g}$), intragranular vortices start to penetrate the grains. It has been proved that the principal dissipation process occurs due to the motion of intergranular vortices. The intergranular vortices are embedded in a diamagnetic grain-medium of effective permeability [23. 26, 27], $\mu_{eff} = f_n + f_s \mu_{cer}$, where

$$\mu_{cer} = \frac{2\lambda}{R_g} \frac{I_1(R_g / \lambda)}{I_0(R_g / \lambda)}. \tag{7}$$

Here R_g is the average grain size approximated by superconducting cylinders of grain aligned along the direction parallel to the applied field. f_n is the fraction of intergranular material, f_s is the fraction of superconducting grains ($f_n + f_s = 1$) and $I_{0,1}$ is the modified Bessel function of the first kind. λ is the London penetration depth of the grains (neglecting grain anisotropy) associated with the screening currents. It is often assumed that λ varies with temperature and field [28]. This relation is given by,

$$\lambda(T, H) = \lambda(0)\left(1 - \frac{H}{H_{c2,g}}\right)^{-1/2}\left(1 - \left(\frac{T}{T_{c,g}}\right)^4\right)^{-1/2}. \tag{8}$$

In the model calculations for the low applied field, the field dependence can be neglected ($H << H_{c2,g}$) and only temperature-dependent London penetration depth is considered. Many measurements have indicated that the measured magnetization is composed of the intergrain and intragrain contributions. Thus the magnetization of granular superconductors can be expressed as,

$$M = f_g M_g + (1 - f_g) M_m, \tag{9}$$

where M_g is the magnetization of the diamagnetic grains and M_m is the contribution to the magnetization from the matrix. f_g given by $1-\mu_{eff}$ is an effective volume fraction of grains. The M_g and M_m can be calculated by using Anderson-Kim model for different parameters. For example, the local intergranular magnetic field, B_m, is determined by the critical-state equation

$$\left|\frac{dB_m}{dx}\right| = J_{cm}, \tag{10}$$

where J_{cm} is the intergranular critical current density. The transparent expression for the intergranular critical current density can be obtained as $J_{cm}= p_m^2 B_{om}/(2a)$ by using $p_m= (2k_m a)^{1/2} / B_{om}$ with $k_m= J_{cm} B_{om}$. The intergranular magnetization, M_m, can be calculated by using Eq. 4. The procedure described above can also be used to calculate intragranular magnetization, M_g, by employing Anderson-Kim model with the temperature dependence of intragranular critical current density assumed as $J_{cg}=J_{cg}(0)[1-(T/T_{cg})^2]^2$[15]. The experimental measurement of AC magnetization can be obtained in terms of the intergranular and intragranular regions by using Eq. (9). On the other hand, the measured susceptibilities χ_n' and χ_n'', including the intergranular and intragranular contributions, are given by

$$\chi_n'= f_g \chi_{n,g}' + (1-f_g) \chi_{n,m}'$$

$$\chi_n''=f_g \chi_{n,g}''+ (1-f_g) \chi_{n,m}''. \tag{11}$$

Here $\chi_{n,m}'$ and $\chi_{n,m}''$ are susceptibilities relating to intergranular regime calculated from Eq. (5) by using M_m. $\chi_{n,g}'$ and $\chi_{n,g}''$ are susceptibilities relating to intragranular regime calculated from Eq. (5) by using M_g. Typical results of model calculation for the AC magnetization and susceptibilities in granular superconductors are shown in Fig. 5. As shown in the Fig. 5(a), the AC magnetization loops as a function of magnetic field at a constant temperature are composed of two sets of hysteresis which can be identified as contributions from the intergrain and intragrain regions. The total AC magnetization is given in the inset of Fig 5(a). The fundamental susceptibilities as a function of temperature for a constant magnetic field are plotted in the Fig. 5(b) by separating intergranular and intragranular contributions. The total fundamental susceptibility is given in the inset of Fig 5(a). The curves of fundamental susceptibility for granular superconductor display two step processes, which reflect the flux penetration between and into the grains, as T decreases. Granular superconductors exhibit two critical temperatures. One is intrinsic (T_{cg}) to the superconductor and the other is characteristic of the coupling between grains (T_{cm}).

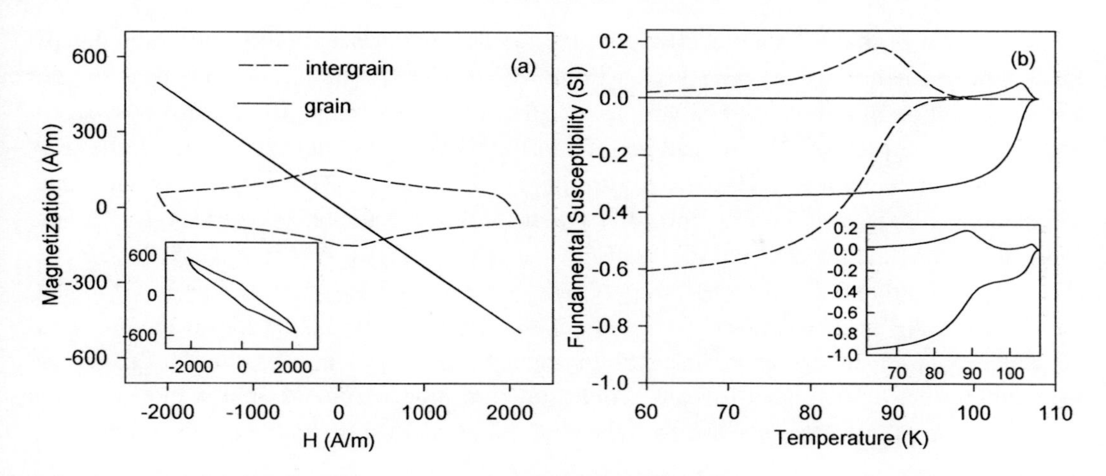

Figure 5 a) Calculated magnetization as a function field at a constant temperature for intergrain and grain contributions. *Inset*: The magnetization, which is sum of these contributions. b) Fundamental susceptibility as a function of temperature at a constant field for intergrain and grain. *Inset*: The susceptibility, which is sum of these contributions.

The prediction of the model for fundamental susceptibilities depends on the volume fraction and grain size. In Figure 6(a) we report the fraction of superconducting grains (f_s) dependence of χ_1' and χ_1'' for R_g= 2 μm whereas Fig 6(b) shows the R_g dependence of χ_1' and χ_1'' for f_s= 0.4. As can be clearly seen from Fig 6(a) and (b), the height of the plateau of χ_1' depends on the fraction of superconducting grains and the radius of the superconducting grains. As f_s is increased, the height of peak in χ_1'' for the intergranular regimes decreases and the breadth in χ_1'' decreases. However the peak height of χ_1'' for the granular regimes increases. On the other hand, the same behavior for the fundamental susceptibilities observed with increasing R_g.

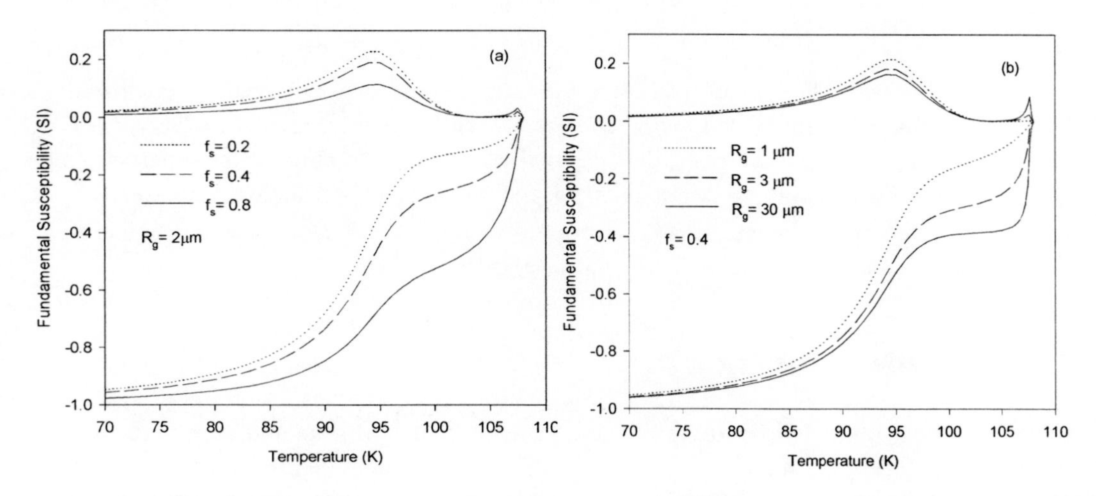

Figure 6 Fundamental susceptibilities as a function of temperature at H_{ac}= 50 A/m, a) for different fractions of superconducting grains, f_s, b) for various grains size, R_g.

The effect of the AC magnetic field on the fundamental susceptibility for a granular Bi-2223 superconductor can be clearly seen from Fig. 7(a) where the measurements of χ_1' and χ_1'' as a function of temperature at various AC fields (H_{ac}= 8, 80 and 160 A/m) with H_{dc}=0 and f=80 Hz are plotted [21]. As can be seen in Fig. 7(a), as AC field is increased, the peak height in χ_1'' increases, and the breadth in χ_1' increases. The corresponding model calculations are plotted in Fig 7(a). The values for the parameters used in this model calculation for intergranular regime are $\lambda(T)$ = 0.2 $[1-(T/T_{cg})^4]^{-1/2}$ μm, $H_{pm}(T)$=9.9 10^3 $[1-(T/T_{cm})]^{1.9}$ A/m, p_m= 0.7 and T_{cm}=103K. $H_{pg}(T)$= 1.4 10^6 $[1-(T/T_{cg})]^2$ A/m, T_{cg}= 106, f_s= 0.34, R_g= 2 μm and p_g= 20 are taken for intragranular regime. Figure 7(b) shows the AC field dependence of the fundamental susceptibility as a function of temperature as a function of temperature that is calculated by using Anderson-Kim model. As AC field is increased, the peak height in χ_1'' slightly decreases and the peak temperature shifts to lower temperature.

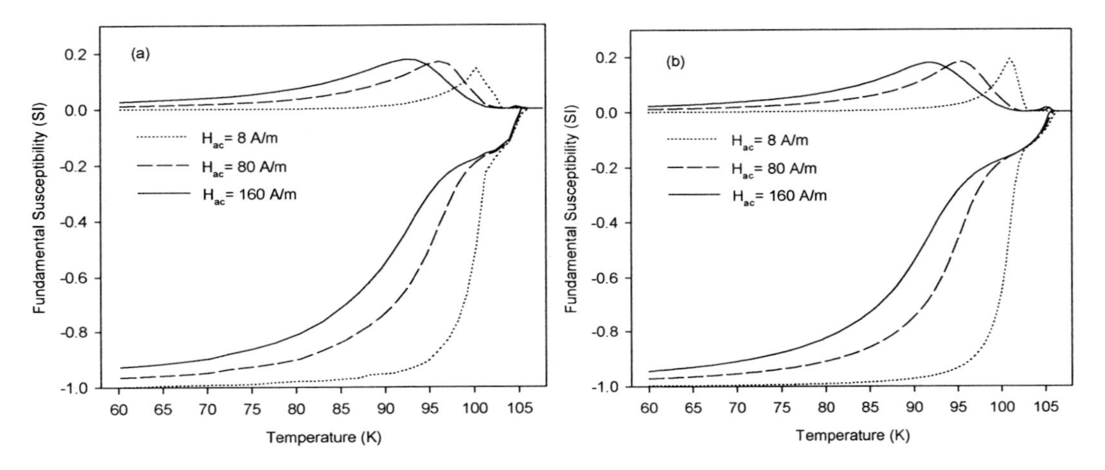

Figure 7. Fundamental susceptibilities as a function of temperature for different AC field amplitudes a) Measurement, b) Calculation

The prediction of the model, including the intergranular and intragranular contributions, for fundamental susceptibilities was consistent with experimental data very well by considering the temperature-dependent effective volume fraction. Nevertheless, some discrepancies between the experimental data and the model calculations are still present, especially in the peak behavior in χ_1''. In order to interpret the AC field dependence in χ_1'', it is necessary to consider the magnetic relaxation effects.

4 Magnetic Relaxation

In the critical-state model the virtually static hysteretic internal field and current distribution are ascribed to the pinning of the flux distribution against the electromagnetic driven force. Thermal fluctuations are known to be a very important factor affecting the critical current density and pinning of vortex structure in high temperature superconductor for which the observed rapid relaxation of magnetization leads to essential dependence of irreversible magnetic properties on induced electrical fields. The basic idea is that the magnetic relaxation arises from thermal activation of flux line or flux bundles over average activation energy barrier. This same effect leads to a reduction of the apparent critical current density comparing with an ideal J_c. In order to explain the magnetic relaxation associated with the decay of persistent current, Anderson [3, 6] proposed the model by considering linear dependence of the pinning potential U on the current density J, namely "giant" flux creep theory based on thermally activated flux motion. In a first approximation, the net barrier is reduced linearly with the current J, according to $U_{eff}= U_o(1- J/J_c)$, where U_o is an average value for activation energy at low current density and J_c is the current density that reduces U_{eff} to zero. Combining this equation with the Arrhenius relation, this leads to the classic equation of flux creep as,

$$J=J_c\left[1-\frac{k_B T}{U_0}\ln(\frac{t}{t_o})\right], \tag{12}$$

where t_o is an effective attempt time and k_B is the Boltzman constant. According to Eq. (12), the relaxation rate depends on the limiting critical current density J_c, the barrier height U_o, and the effective attempt time t_o. Later, starting with Beasley *et al* [29], several authors proposed the models by considering nonlinear $U(j)$ dependencies. More recent different creep models, nonlinear relation between U and J, have been proposed such as the Vortex Glass (VG) [30] and Collective Creep (CC) [31, 32] with an activation energy barrier U depending explicitly on current density J. Both the CC and VG models are based on elastic vortex motion. The general form of the current dependence of the activation energy, U_j, is given by,

$$U_j=\frac{U_o}{\mu}\left[\left(\frac{J_{co}}{J_c}\right)^{\mu}-1\right], \tag{13}$$

where μ is an important exponent to characterize vortex dynamics. Many different values have been predicted theoretically for the parameter μ. In the collective creep theory there are three different regimes: single vortex, small and large bundles characterized by μ=1/7, 3/2 and 7/9 respectively. The giant flux creep model [3, 6] case corresponds to μ= -1 and the Zeldov model's [33] form to $\mu \to 0$ as special cases. The theory of collective creep has been reviewed by Blatter *et al.* [34]. According to them, the form for the critical current density, J_c, is given by,

$$J_c(T, B, f)=J_{co}(T,B)\left[1+\frac{\mu k_B T}{U_0}\ln(\frac{f_0}{f})\right]^{-1/\mu}, \tag{14}$$

where f_o is an attempt and f is the measurement frequency. The temperature dependence of the critical current density and the apparent activation energy are chosen according to the δl type collective pinning model [35]. For the magnetic field dependence of the current density, we have chosen the form as $J_{co}(B)= J_{co}(0) / (1+B/B_0)$ in agreement with Anderson-Kim model. Generally, the field dependence activation energy can be described by a power law. In this work, in order to avoid introducing another parameter and for simplicity, the field dependence of U is ignored. Therefore, in the model calculations, the temperature and field dependencies of the critical current density and activation energy have been used as,

$$J_{co}(T,B)=\left[1+\left(\frac{T}{T_c}\right)^2\right]^{-1/2}\left[1-\left(\frac{T}{T_c}\right)^2\right]^{5/2}\frac{J_{co}(0)}{1+B/B_0(T)}, \tag{15}$$

$$U_0(T)=U_0(0)\left[1-\left(\frac{T}{T_c}\right)^4\right], \tag{16}$$

where B is local magnetic field and $B_0(T)$ is a temperature-dependent parameter assumed as $B_0(T)=B_0(0)(1-T/T_c)^{3/2}$. Such temperature dependencies of J_{co} and U_0 arise within the collective pinning models, where the vortices are supposed to be pinned by randomly distributed weak pinning centers. When Eq.(14) is combined with Eqs.(15) and (16), the temperature, field and frequency dependencies of current density can be obtained. The magnetization can be calculated by using critical-state equation for the applied time-dependent external magnetic field $H(t)=H_{dc}+H_{ac}\cos(wt)$, where $w=2\pi f$. The complex AC susceptibility, $\chi_n=\chi_n'+i\chi_n''$, can be derived by means of the Fourier transform.

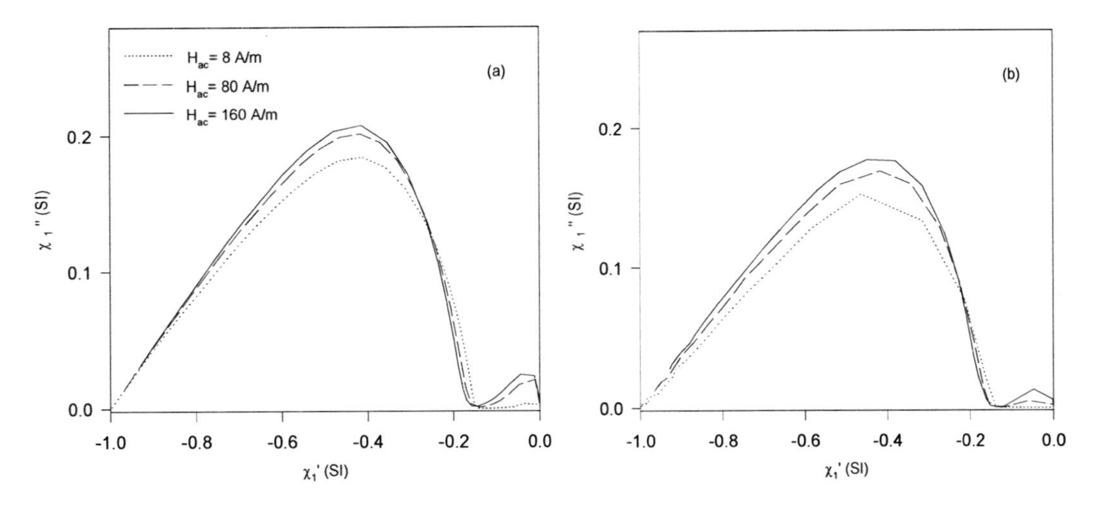

Figure 8 The Cole-Cole plots of fundamental susceptibilities depend on AC field amplitude, a) Calculated, b) Measured

The model calculations have been performed using the procedure described above and the parameters used here are $\lambda=0.2$ μm, $f_0=10^5$ Hz, $U_0(0)=0.6$ eV, $J_{co}(0)=7.8\times10^6$ A/m^2, $B_0(0)=3200$ A/m, $J_{cg}(0)=4.2\times10^9$ A/m^2 $f_s=0.31$ and $R_g=2$ μm. The Cole-Cole plots of the calculated fundamental susceptibilities are plotted in Fig. 8(a) for different AC fields (H_{ac}=8, 80 and 160 A/m) at a fixed frequency (f=80 Hz). As can be clearly seen from Fig. 8(a), when AC field is increased, the χ_1'' peak increases unlike that in Fig. 7(b). This is because the present model includes the effect of thermally activated intergranular regime, which results in a significant decrease of the pinning force density close to T_{cm}. The Cole-Cole plots of measurements fundamental susceptibilities obtained from Fig. 7(a) are given in Fig. 8(b). Comparing Fig.8(a) with Fig.8(b), it can be seen the similar behavior in both of them.

The magnetic behavior of HTS in the mixed state is typically nonlinear. For this reason, the study of the higher harmonics of the AC susceptibility represents an interesting tool to investigate the pinning characteristics of the superconducting materials. In general, the analysis of the third harmonics of the AC susceptibility can be a very powerful tool to discriminate between the different kinds of flux dynamics governing from the magnetic behavior of high temperature superconductors. The simple critical-state models account for the generation higher harmonics. Nevertheless these models do not explain experimental observations where the peak heights of the harmonics strongly depend on observed external parameters (AC field amplitude, frequency and etc.). Recently, the magnetic relaxation effects have been considered to explain the external parameter dependence of harmonic susceptibilities [37-41].

The effect of the AC magnetic field on the third harmonic susceptibility can be seen from Fig. 9 where the calculated χ_3' and χ_3'' as a function of temperature for different AC fields (H_{ac}= 8, 80, 160, 320, 640 and 1000 A/m) with zero DC magnetic field at a fixed frequency (f= 80 Hz) are plotted. Figure 9(a) shows that the real part of third harmonic oscillates between positive and negative values. The ratio in absolute value between the positive and negative peaks changes with increasing applied field. In other words, as the ac field is increased, the positive peak decreases, the small negative peak increases and becomes higher than the negative peak and the peak temperature shifts to lower temperature. As can be seen in Fig 9(b), the imaginary component of third harmonic is positive at higher magnetic field. However, when the ac field is decreased, the positive peak decreases and the peak temperature shifts to higher temperature, the additional small negative peak occurs at near T_{cm}. The Cole-Cole plots of third harmonic are shown in Fig. 9(c). As AC field amplitude is increased, the polar plots turn in direction of anticlockwise and tend to occupy the left semi-plane. The predictions of the model for the field and temperature dependencies of the third harmonics were found to be in good agreement with experimental data [41].

On the other hand, the model incorporating both the concepts of CC and critical-state has been also considered to explain frequency dependence of the ac magnetic field of the harmonic susceptibilities. The measured third harmonics as a function of temperature at H_m=400 A/m for various frequencies (f=80, 1100 and 2110 Hz) for a melt textured YBCO are plotted in Fig. 10(a) [42]. Third harmonic measurements appear pronounced frequency dependence. The polar plots shown in Fig. 10(a) turn in the direction of clockwise when frequency is decreased.

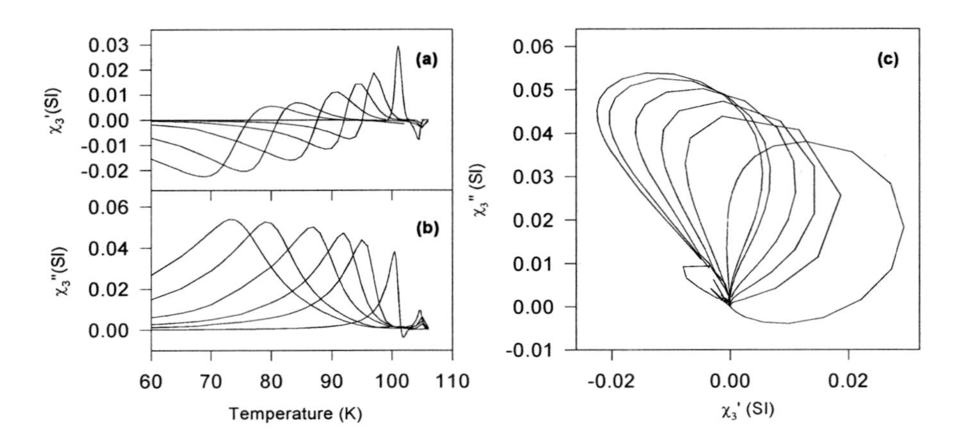

Figure 9. Calculated third harmonic susceptibilities for different AC fields. AC field amplitudes are 8, 80, 160, 320, 640 and 1000 A/m for the curves shown from right to left. a) Real components versus temperature, b) Imaginer components versus temperature, c) Cole-Cole plots

Corresponding model calculations for Cole-Cole plots are plotted in Fig. 10(b). The values for the parameters used in this model calculation have been given in Ref. [42]. The model calculations show that a change of the frequency not only affects the position of peak in χ_3' and χ_3'' but also modifies the shape of the third harmonics. Comparing Fig. 10 (a) with (b), model calculations are in good agreement with the experimental results.

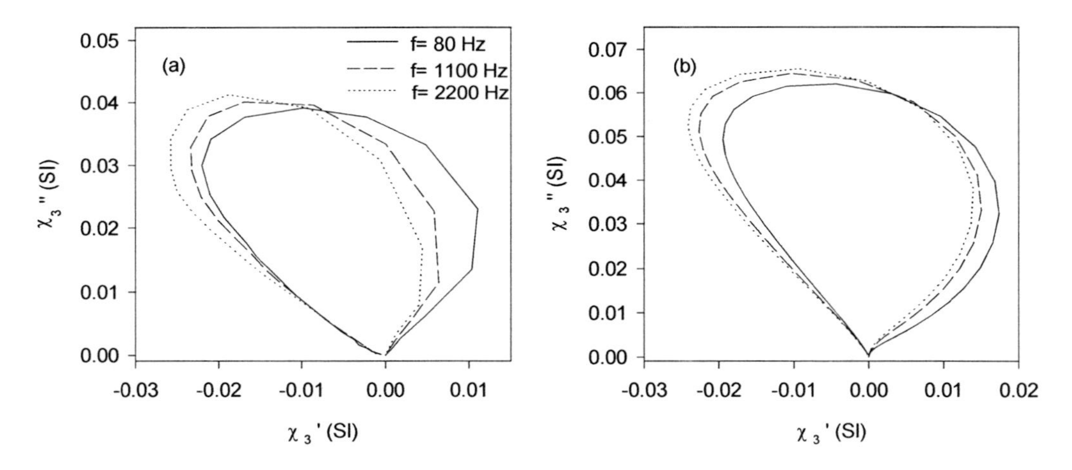

Figure 10 Cole-Cole plots of third harmonics for H_m=400 A/m and f = 80-2110 Hz, *a*) Measured, *b*) Calculated.

5 Conclusion

In this chapter, we have briefly reviewed the AC magnetic susceptibility technique and critical-state models. The first and third harmonics of the AC magnetic susceptibilities for high temperature superconductors have been investigated. The experimental AC susceptibility data has been compared with several theoretical models. It is found good agreement between experimental data and the numerical results obtained in the basis of a theoretical model that incorporates collective creep concepts in a critical-state equation. The following results were obtained from this model calculation. I) As the AC field amplitude is increased, the peak temperature in χ_1'' shifts to lower temperature and its peak increases. II) The same effect is observed for frequency dependent. Nevertheless the peak temperature in χ_1'' shifts to higher temperature as the frequency is increased. III) The shape of the Cole-Cole plots of third harmonic changes with the variation of AC field amplitude and frequency. If AC field amplitude is decreased, the polar plots turn in the direction of clockwise and tend to occupy the right semi-plane. The same behavior is obtained for frequency dependent.

Acknowledgement

I wish to thank to Prof. N. Navruz for a critical reading and encouragement.

References

[1] Bean, C. P. *Rev. Phys. Rev. Lett.* 1962, 8, 250.
[2] Bean, C. P. Rev. Mod. Phys. 1964, 36, 31.
[3] Anderson, P. W. *Phys. Rev. Lett.* 1962, 9, 309.
[4] Kim, Y. B.; Hempstead, C.F.; Strmad, A.R. *Rev. Mod. Phys.* 1962, 36, 43.
[5] Kim, Y. B.; Hempstead, C.F.; Strmad, A.R. *Phys. Rev.* 1963, 129, 428.
[6] Anderson, P. W.; Kim, Y.B. *Rev. Mod. Phys.* 1964, 36, 39.
[7] Fietz, W. A.; Beasley, M. R.; Silcox, J.; Webb, W. W. *Phys. Rev.* 1964, 136, A355.
[8] Watson, J. H. V.; *J. Appl. Phys*, 1968, 39, 3406.
[9] Irie, F.; Yamafuji, K. *J. Phys. Soc. Jpn.* 1967, 23, 255.
[10] Ji, L.; Sohn, R. H.; Spalding, G. C.; Lobb, C. J.; Tinkham, M. *Phys. Rev. B.* 1989, 40, 10936.
[11] Chen, D. X.; Goldfarb, R. B. *J. Appl. Phys.* 1989, 66, 2489.
[12] Ishida, T.; Goldfarb, R. B. *Phys. Rev. B.* 1990, 41, 8939.
[13] Hulbert, J. A. *Brit. J. Appl. Phys.* 1965, 16, 1657.
[14] Yasuoka, H.; Tochihara, S.; Mashino, M.; Mazaki, H. *Physica C.* 1998, 305, 125.
[15] Müller, K. H. *Physica C.* 1989, 159, 717.
[16] Ozogul, O.; Aydınuraz, A. *Supercond. Sci. Technol.* 2001, 14, 184.
[17] Geshkenbein, V. B.; Vinokur, V. M.; Fehrenbacher, R. *Phys. Rev. B.* 1991, 43, 3748.
[18] Clem, J. R.; Bunmble, B.; Raider, S. I.; Gallagher, W. G.; Shih, Y. C. *Phys. Rev. B.* 1987, 35, 6637.
[19] Savvides, N. *Physica C.* 1990, 165, 371.
[20] Q.H. Lam, Q. H.; Jeffries, C. D. *Physica C.* 1992, 194, 47.
[21] Ozogul, O. J. *Supercond.* 2004, 17, 513.
[22] C.Y. Lee, C. Y.; Kao, Y. H. *Physica C.* 1995, 241, 167.
[23] Clem, J. R. *Physica C.* 1988,153-155, 50.
[24] Dersch, H.; Blatter, G. *Phys. Rev. B.* 1988. 38, 11391.
[25] Senoussi, S.; Oussena, M.; Ribault, M. *Phys. Rev. B.* 1987, 36, 4003.
[26] Müller, K. H.; Macfarlanc, J. C.; Driver, R. *Physica C.* 1989, 158, 69.

[27] Angurel, L. A.; Lera, F.; Rillo, C.; Navarro, R. *Cryogenics*. 1993, 33, 314.
[28] Schowlow, A. L.; Devlin, G. E. *Phys. Rev.* 1959, 113, 120.
[29] Beasley, M. R.; Labusch, R.; Webb, W. W. *Phys. Rev.* 1969, 181, 682.
[30] Fisher, M. P. A.; *Phys. Rev. Lett.* 1989, 62, 1415.
[31] Larkin, A.; Ovchinnikov, Y.N. *J. Low Temp. Phys.* 1979, 34, 409.
[32] Feigel'man, M. V.; Vinokur, V. M. *Phys. Rev. B*. 1990, 41,8989.
[33] Zeldov et al. *Appl. Phys. Lett.* 1990, 56, 680.
[34] Blatter, G.; Feigal'man, M. V.; Geschkenbein, V. B.; Larkin, A.; Vinokur, V.M. *Rev. Mod. Phys.* 1994, 66,1125.
[35] Griessen, R. et al., *Phys. Rev. Lett.* 1994, 72, 1910.
[36] Qin, M. J.; Ong, C. K. *Phys. Rev. B*. 2000, 61, 9786.
[37] Zhang, Y. J.; Qin, M. J.; Ong, C. K. *Physica C*. 2001, 351, 395.
[38] Gioacchino, D. Di.; Tripodi, P.; Celani, F.; Testa, A. M.; Pace, S. *IEEE Trans. Appl. Supercond.* 2001, 11, 3924.
[39] Senatore, C.; Polichetti, M.; Zola, D.; Matteo, T. Di.; Giunchi, G.; Pace, S. *Supercond. Sci. Technol.* 2003, 16, 183.
[40] Adesso, M. G.; Polichetti, M.; Pace, S. *Physica C*. 2004, 401, 196.
[41] Ozogul, O. *Phys. Stat. Sol.* (a). 2005, 202, 1973.
[42] Ozogul, O. J. *Supercond.* 2005, 18, 503.

INDEX

A

B

C

D

E

F

N

O

P

Q

R

S

T

U

V

W

X

Y